QUICK(ER) CALCULATIONS

QUICK(ER) CALCULATIONS

How to Add, Subtract, Multiply, Divide,
Square, and Square Root More Swiftly

Trevor Davis Lipscombe

OXFORD
UNIVERSITY PRESS

Great Clarendon Street, Oxford, OX2 6DP,
United Kingdom

Oxford University Press is a department of the University of Oxford.
It furthers the University's objective of excellence in research, scholarship,
and education by publishing worldwide. Oxford is a registered trade mark of
Oxford University Press in the UK and in certain other countries

Published in the United States of America by Oxford University Press
198 Madison Avenue, New York, NY 10016, United States of America

British Library Cataloguing in Publication Data

Data available

Library of Congress Control Number: 2020948798

ISBN 978–0–19–885265–0

Printed and bound in the UK by
TJ Books Limited

To Kath

*The sum that two married people owe to one another defies calculation. It is an infinite
debt that can only be discharged through all eternity.*
Johann Wolfgang von Goethe, Elective Affinities *Book 1, Chapter 9*

For Mary, Ann, Clare, Therese, and Peter
"Dad watching you do math homework is like Michael Phelps watching you drown."

Preface

Multiplication is vexation
Division is as bad
The rule of three doth puzzle me
And practice drives me mad.

From an Elizabethan manuscript of 1570
(as quoted in *Bartlett's Familiar Quotations*
15th and 125th Anniversary edition
(Boston, MA: Little Brown, 1980), p. 917)

Coffee shops crowd onto every city block, or so it seems. Some discerning souls refuse to drink at such places. Instead, they purchase fair-trade beans, grind them up, and use a French press to produce the life-giving liquid. Others forego the brioche sold by supermarkets, preferring to buy their baked goods from small craft bakeries.

What better way to accompany your hand-crafted coffee and croissant than with artisanal arithmetic? Stow your computer and use the most affordable and ecofriendly calculator you possess—your brain. This book shows you how to carry out simple arithmetic—addition, subtraction, multiplication, division—as well as how to square or to find roots (square, cubic, and even quintic) quickly. Faster, in fact, than you could walk across the room, find a calculator, put in the numbers, and hit the big red button with a C on it. Or use your landline to call your smart phone, find it, and get to the calculator function.

Why bother? I offer two different answers, neither of which is original. First, "For those who believe, no explanation is necessary; for those who do not believe, no explanation is possible," which is variously attributed, in chronological order, to Saint Thomas Aquinas, Saint Ignatius of Loyola, and Franz Werfel—the script-writer for the movie *The Song of Bernadette*. In this context, those who have fun frolicking with formulas need no explanation for why they should attempt to carry out fantastically fast arithmetic.

The second answer belongs to Sir George Mallory, who repeatedly tried to climb Mount Everest, and who eventually died on its slopes. When asked why he made so many attempts, Mallory replied "Because it's there"—a phrase often misattributed to Sir Edmund Hilary, who, with Tenzing Norgay, first conquered the mountain in 1953. In other words, we should practice quick calculations simply because we can. After all, international grand masters play chess not only against other international grand masters, but also against the clock.[1] Likewise, in the cutthroat world of crossword competitions, it's not sufficient merely to complete, say, *The Times* or *The New York Times* puzzle—you have to do so faster than everyone else does. So, why not calculate against the clock? I hope this short book will be a step in that direction. As a further incentive, every 2 years, the *Weltmeisterschaften im Kopfrechnen* (loosely translated as the World Cup in Mental Arithmetic) takes place in Germany. To give a flavor of the competition, events include adding 10 ten-digit numbers, multiplying 2 eight-digit numbers, and calendar calculations (in which contestants peruse a list of dates from 1600–2100 and have 1 minute to state correctly on which day of the week they fell). To provide a benchmark, the current World Record stands at 66 correct answers in the allotted 60 seconds. As for calendrical calculations, see "Interlude III: Doomsday" later on in this book.

Sir Edmund Hillary, together with Charles Lindbergh—the first person to fly solo across the Atlantic—show that rapid mental mathematics can save your life. Hillary, as he reports in his book *High Adventure*, pushed on to the summit of Mount Everest, but knew oxygen was running low. Everest (or Sagarmatha in Nepalese), rises 29,029 feet (8,840 m) above sea level—although China and Nepal dispute the actual height of the mountain, whose summit forms part of their border. Should he and Tenzing continue, or retrace their steps down the mountain? Hillary's quick estimate convinced him to go on. They did so, and claimed the honor of being the first two people to stand atop the world's tallest mountain.

Lindbergh had a similar predicament, requiring rapid aviation arithmetic: did he have enough fuel to make it over the ocean? A wrong answer would have spelled disaster. The 450 gallons (1,704 liters) of fuel in his plane, he reckoned correctly, sufficed to power the *Spirit of St. Louis* from Roosevelt Field, New York (now the home of a shopping mall) to Paris, touching down at the Le Bourget Aerodrome on the evening of May 21, 1927.

---

[1] There's a formula to rate chess prowess, the Elo rating system. If you play N games, win W, lose L, and play opponents whose ratings sum to R, then your rating is $[R + 400\,(W - L)]/N$. As you can't lose more games than you play, your score will increase if you play only those people whose ratings are higher than 400.

Preface

There are practical benefits—less drastic than saving a life—in calculating more quickly. In an era of timed multiple-choice tests, the
quicker you perform a particular mathematical operation, the more
time you can devote to the other questions. That's why some sections
in this book are devoted not to *exact* answers, but to *approximate* ones. If
you can calculate an approximate answer swiftly to within a couple
of percent of the actual answer, that's often sufficient to determine
which multiple-choice answer to select. Or at least that permits you
to eliminate a couple of answers, thereby increasing your chances of
plumping for the right one purely by guesswork.

Isaac Newton wrote that if he had seen further than anyone else, it
was because he had stood on the shoulders of giants. Likewise, mathematical techniques to calculate quickly are not recent inventions. They
come from a variety of cultures and from across the globe. Western
scholars only recently began to pay serious attention to the mathematics of other cultures, giving rise to the subject of ethnomathematics.
Likewise, the history of mathematics reveals many different ways that
non-European cultures devised to multiply, divide, and extract square
roots, some of which are included in this slim volume. I am privileged
to give a glimpse into how other cultures have calculated, and hope
some readers may delve further into the topic. There are some fresh
techniques original to this book, however, such as the section on multiplying and dividing by irrational numbers and the entries on squaring
certain three-digit numbers.

I hope that *Quick(er) Calculations* achieves what the BBC sets out to
do: inform, educate, and entertain. By the end, I trust that you will be
able to compute answers to basic arithmetical problems far faster than
you have up to this point. Being able to do so may be of great help for
those still in high school, studying mathematics, physics, or chemistry.
It should also help those who have already entered into professions, who
are card-carrying scientists, engineers, accountants, and actuaries. I also
hope the book will familiarize you with, and give you more appreciation
for, those of different times and places who have devised ingenious ways
to calculate answers to questions we still pose today.

Trevor Lipscombe
Baltimore, Maryland
Feast of the Assumption of the Blessed Virgin Mary, 2020

Acknowledgments

Another damned thick square book! Always scribble, scribble, scribble! Eh! Mr. Gibbon.

> William Henry, Duke of Gloucester,
> after receiving volume II of Gibbon's
> *Decline and Fall of the Roman Empire*

My family is usually pleasant and peaceful. Except during the card game "Pounce." Intense competition takes control and, with the round completed, I'm invariably in last place. Something odd happens, though, when we tot up the scores. Someone says, "Add on 30, take off 4," or something like it, poking gentle fun at the way I do math. One night, after being drubbed yet again, the idea for this book pounced (bad pun intended) upon me.

I owe thanks, then, to Quick Hands McGee, the Evens, and the Odds (you know who you are) for serving as my inspiration. I also thank them for all the times we've spent over the years enduring that most unenjoyable of things, mathematics homework. It's there that my kids suffered while I did the whole "Take off 60, add on 3" routine, and they put up with my oft-vented frustration that no-one teaches elegant mathematics in grade school, middle school, high school. . . .

My own high-school mathematics teachers deserve a great deal of thanks. John Evans and Ken Thomas of the Hugh Christie School in Tonbridge, Kent, were always encouraging and helped steer me successfully down the road to university. I hope this book does them, and the school, proud, and helps their current and future students do arithmetic more quickly.

My school report, from when I was 10 years old, asserts "Trevor must be much more careful in Mechanical Arithmetic. He makes more careless errors than he should." It turns out that what Miss Watson wrote at Icknield Primary School in Luton several decades ago remains true today. I am especially thankful, then, for Camille Bramall's attentive copyediting; she has graciously rooted out more careless errors than I should ever have committed. Those that remain are my fault alone, serving as sure signs I should have paid more attention to what Miss Watson was trying to teach me.

My gratitude also extends to those laboring in the vineyards of Oxford University Press, in particular Sonke Adlung and Katherine Ward, for seeing something of merit in my efforts. I hope you will, too.

Contents

Introduction

The different branches of arithmetic—ambition, distraction, uglification, and derision.
The Mock Turtle, *Alice's Adventures in Wonderland*, by Lewis Carroll.

My mother—a formidable Londoner who lived through the Blitz and who would delight in being described as a No Nonsense Person—had an unusual way of telling time. If the clock showed 6:25, she proclaimed it "Five and twenty past six." Ten minutes later, naturally, she reported the time as "Five and twenty to seven."

My mum's way to report time shows an insight. Namely, sometimes it's easier to break a calculation up into two steps, rather than one. Rather than adding 25, instead add 20, then add 5. Dissolving a difficult problem into two simpler ones lies at the core of accelerating your arithmetic. Two other key ingredients—which, depending on your viewpoint, are either trivial and obvious or deeply profound—is that you can add 0 to any sum without doing any harm, or multiply any expression by 1 without changing the outcome. Enigmatically, much depends on what the 0 and the 1 happen to be. When presented with a complicated sum, it may be easier to add 3 to it, to create a far simpler addition, and then subtract 3 at the end. As $3 - 3 = 0$, you have added 0—and thus have the same answer—but have made the calculation quicker. Likewise, for some divisions, multiplying the top by 4 may make life easier and, if you divide by 4 later on, you will have multiplied by 4/4, which is 1, leaving the answer unchanged. All shall be revealed.

This book begins with a brief glance at how you may calculate currently, and dispenses advice on how to improve both speed and accuracy. Then comes a chapter devoted to additions and subtractions, before turning to the accounting challenge of summing up long columns of numbers at speed. From this follows a bird-by-bird guide, so to speak, to multiplication and division by specific numbers. Some of the tips and techniques hold for more than one number so, if you read straight through that section, there will be some repetition. This is good for making a point, but it also means that at any time you wish to learn a technique for a specific number, you can simply go to the number itself and see what to do. The next division (pun intended), deals with non-specific numbers. Things such as "multiplying two numbers, both of which are just over 100." These lend themselves to impressing friends, relatives, students, or even an audience ("give me two numbers, any two numbers, between 100 and 110. . ."). Practical applications for square roots abound, especially if you know of Pythagoras' theorem. This gives rise, then, to a chapter devoted to finding the square root, approximately, of a given number. This section also deals with another way to impress people, in that they can use a calculator to square, cube, or raise to the fifth power any two-digit number they choose, and you can divine that original number in a matter of seconds using only your brain—often simply by inspection.

The book closes with some approximate ways to multiply and divide. For example, to divide by 17, you can get fairly close (within 2 percent) by multiplying the number by 6 and dividing by 100. There are also a number of ways to calculate multiples of $\sqrt{2}$, the Golden Ratio, e, and π. For more details, read on!

Scattered within these sections are interludes. These are digressions, hopefully enjoyable, that touch lightly on the more technical material in the book. They deal with supermarket check outs, determining what day of the week a particular date fell on (useful for those mulling a run at the World Cup in Mental Arithmetic), and a look at calculations through non-modern or non-Western eyes. The interludes also present a mathematical trick involving the number 111,111 and—always important—the ways to estimate the number of extra-terrestrial lifeforms.

The challenge in writing a book on how to add, subtract, multiply, divide, and find square roots is to do so in a way that it doesn't end up with the thrilling prose found in rental agreements or the instructions that come with do-it-yourself furniture. Peppered within each chapter, then, are a number of fanciful "applications." These include the number of people who accompanied King Henry VIII to France and took part in one of history's greatest binges, the Field of the Cloth of Gold; the amount of bread consumed annually by the court of King Charles I; the number of goals scored by Charlton Athletic during their years in the Premier League; and the number of lines of poetry in Shakespeare's sonnets. While these are far-fetched, you can apply the methods in the book to calculate the answers. Arguably more important, they show that you can find opportunities, or excuses, to use quick(er) calculations almost everywhere.

Challenge

The aim of this book is to help you calculate more quickly. To begin, I invite you to rise to this challenge. Try the 25 questions below (26 if you feel up to it), before reading the book, and see how long it takes. Once you've read the book and learned the streamlined ways to calculate, try them again to see how many seconds, or minutes, you have shaved from your time. To help, sections of the book that may later assist you are given in parentheses.

Pencil and brain sharp? Go!

1. $171 - 46$ (Subtract by adding)
2. $1.27 + 0.98$ (Two-step subtraction)
3. The birthdays of my family fall on the following days of the month. Add them: 14, 17, 2, 23, 13, 20, 27 (Tallying in tens)
4. The first 10 Fibonacci numbers are 1, 1, 2, 3, 5, 8, 13, 21, 34, 55. Add them (Adding numbers mystically)
5. One of my kids, preparing to run a marathon, kept a "jog log" to track her miles. Over a 10-day period, she ran the following number of miles each day: 8, 3, 2, 3, 9.5, 4, 4, 3, 2, 3. How many miles did she run in total? (Cancel as you calculate)
6. 87×1.1 (Multiply by 11)
7. $4.1 \times 1,800$ (Multiply by 18)
8. 53×2.9 (Multiply by 19, 29, 39. . .)
9. $2,100 \times 4.3$ (Multiply by 21, 31, 41. . .)
10. $3,018/7.5$ (Multiply or divide by 75)
11. 620×11.1 (Multiply by 111)
12. 9.5×130 (Multiply by $316\frac{2}{3}$, $633\frac{1}{3}$, and 950)
13. 8.5×75 (Multiply two numbers, 10 apart, ending in 5)
14. 430×0.98 (Multiply a one- or two-digit number, less than 50, by 98)
15. 27×2.9 (Multiply two numbers that differ by 2, 4, 6, or 20)
16. $2,300 \times 2.7$ (Multiply two "kindred" numbers)
17. 940×9.7 (Multiply two numbers just under 100)
18. 107^2 (Multiply two numbers (or square a number) just over 100)
19. 79^2 (Square any two-digit number ending in 1 or 9)

20. 4.5^2 (Square any number ending in 5)
21. 39.1^2 (Square any three-digit number with middle digit 0 or 9)
22. What, approximately, is $\sqrt{7.9}$? (Square roots—any style!)
23. What is the cube root of 175,616?
(Find the cube root of a mystery perfect cube)
24. What, roughly, is $130/1.7$? (Multiply or divide by 17)
25. What, approximately, is 8π? (Multiply by π)
26. (Bonus) January 3, 1957 was a Thursday. The space race began on October 4, 1957, when the Soviet Union launched the satellite Sputnik. Which day of the week was that?
(Appendix I Calculating Doomsday)

1

Arithmetical Advice

2 + 2 in #Theology can make 5
> A tweet on January 5, 2017 by Fr. Antonio Spadaro,
> SJ, a close friend of Pope Francis, causing much head
> scratching among mathematicians and theologians

In the end the Party would announce that two and two made
five, and you would have to believe it. It was inevitable that they
should make that claim sooner or later: the logic of their position
demanded it.

GEORGE ORWELL, 1984

Before delving into the delights of quicker calculations, there are a few
pointers worth following. These will help speed things up and reduce
mistakes.

Take less time on your times table

Consider $7 \times 9 = 63$.

How did you read it? As youngsters, we often learn to recite our
multiplication tables along with other kids in the class. If so, you may
have chanted "seven times nine equals sixty-three." True—but slow.[1]
Shred seconds from the time it takes to tell your times table by replacing
this laborious process with a question. "Seven nines? Sixty-three." This
provides a golden opportunity to refresh your acquaintance with the
multiplication tables, but in this new format. There's an added bonus:
Not only does the new way shave two syllables off the phrase, it's also,
well, a new way to say it. In other words, you don't have to repeat
your multiplication tables in the slow sing-song way that you may have

---

[1] An early English arithmetic book from 1552, the anonymous *An Introduction for to
Lerne to recken with the pen or wyth the counters accordynge to the trewe cast of Algorisme* tells us "7 tyme
9 maketh 63," so slow-paced multiplication is nothing new. The book, though, went
through eight editions, making it a sixteenth-century mathematical bestseller.

learned. Recite your seven times table your usual way, but time yourself. Then try the compact way, saying as swiftly as you can, "One seven? Seven. Two sevens? Fourteen. . ." Better yet, time how long it takes you to go all the way from "One times one is one" up to "Twelve times twelve is one hundred and forty-four" and compare with the time it takes to go from "Once one? One" up to "Twelve twelves? One hundred and forty-four!" See how much time you can save!

The case of the crooked columns

Your handwriting is precisely that: yours. Even if it is unintelligible to other readers, a problem that plagues medical doctors, that won't matter when you try to do quick calculations, unless you have some numbers that can be easily mistaken for others.

A bigger point, though, is keeping the tens and the units columns aligned. It is all too easy to write down a sum, say, as:

$$3\ 4+$$
$$\underline{17}$$

In which the 4 drifts dangerously to the right, or, viewed differently, the 7 crowds out the 1. Over the course of the calculation, especially if done at speed, things can get messed up and suddenly you are calculating:

$$304+$$
$$\underline{170}$$

Putting zeros where the sort-of blank spaces were. Instead of getting the result 51, the answer on offer is 474. So, please, keep columns aligned. This may seem like no big deal, and when you add just two numbers of two digits, it probably isn't. But if you have to add up long lists of numbers that have multiple digits, mistakes can rack up rapidly.

Welsh mathematician Robert Recorde (ca.1512–1558) wrote some of the first books in English on arithmetic. In *The Ground of Artes*, published in 1543, Recorde advises students engaged in currency calculations: "If your denominations be pounds, shillings, & pennies, write pounds under pounds, shillings under shillings, and pennies under pennies: And not shillings under pennies, nor pennies under pounds."[2] Authors

---

[2] "Yf your denominations be poundes, shyllyns, & pennes, wryte poundes under po[ndes], shyllynges under shyllynges, and pennys under pe[n]nys: And not shyllynges under pennys, nor pe[n]nys under poundes."

of arithmetic books have been urging readers to keep columns aligned for at least 550 years!

Guess what? Make estimates

As your mental math heats up, the time to calculate quantities drops impressively. The ability to mess up, however, increases—unless you make estimates. Guessing an answer takes a fraction of a second but can catch embarrassing mistakes. In the previous example, if you start by observing that 34 is roughly 30 and 17 is almost 20, you'd expect an answer of about $30 + 20 = 50$. If you get the answer 51, then you'd feel happy and confident. Mix up your columns and get 474 and your rough estimate of 50 screams at you to go back and check.

There is a subtle art in estimation, which is perhaps not quite so subtle. In an addition, such as $34 + 17$, it helps to overestimate one number (say 20 for 17) and compensate for that by underestimating the other number (30 for 34). That way, you end up with a rapid estimate that's likely to be fairly close to the real answer. For something more complicated, such as $513/3.78$, you might bracket the answer with two estimates. For a fraction, you can overestimate the top (the numerator) and underestimate the bottom (the denominator), both of which overestimate the fraction. Thus, $513/3.78 < 540/3$ and so $513/3.78 < 180$. Then do the reverse—underestimate the numerator, overestimate the denominator, and so underestimate the answer. Here we could set $513/3.78 > 500/4 > 125$. Whatever answer we get is only reasonable if it lies between 125 and 180.

What's the point? Extra zeros and decimals

Another vital tip to beef up your calculation is to ignore inconvenience—such as zeros at the end of numbers, or decimals in the middle. These can be stripped out, so that they don't get in the way of calculation, and then reinserted later on. After all, you have already estimated the answer—so you know how many zeros to add or where to insert the final decimal point. There's no need to slow yourself down by keeping track of all those decimal points or extra zeros at each step of the journey.

Consider, for example, 0.9×1.2. In regular mode, you would write these numbers one above the other, start multiplying with the decimals in place to come up with an answer. Don't. Let critics turn red with anger, white with fear, or watch you and go green with envy.

First, make an estimate: 0.9 is roughly 1. So, too, is 1.2. We expect our answer to be about $1 \times 1 = 1$. Remove the decimal points to leave 9×12, whose answer you've known since your early days, since $9 \times 12 = 108$. As the answer is about 1, stick the decimal point in to get the actual answer, namely $0.9 \times 1.2 = 1.08$.

As a second example, think of $1,300 \times 40$. As before, estimate. We know 1,300 is more than 1,000, so the answer is going to be a bit more than $1,000 \times 40 = 40,000$. Now 13×4—as every card player knows— is 52. And so, as the answer slightly exceeds 40,000, we must have $1,300 \times 40 = 52,000$.

Let me make an appeal: clothe the naked decimal point! It's bad style to write one tenth as .1, the reason being that it is too easy to overlook the decimal point. No-one ever writes 01, so if you always write 0.1, you know it's one tenth, and you are unlikely to mistake it for 01.

And, now our decimal points are clothed in all their splendor, here's a reminder about scientific notation, which won't be used much in this book, I admit. Rather than write 12,345, for example, you can write it as 1.2345×10^4. If you struggle to remember the power of 10 that's used, imagine a decimal point being inserted, which would give 12,345.0. How many numbers are there before the decimal point? Five. Subtract 1, to get 4, so $12,345 = 1.2345 \times 10^4$. But now let's suppose we have the number 0.054321. The first non-zero number after the decimal point is a 5, which occurs two places in. Hence, we write this as 5.4321×10^{-2}. If you feel the need, $12,345 \times 0.054321 = 1.2345 \times 5.4321 \times 10^2$, for when you multiply powers of 10 together, you add the exponents. As $12 \times 5.5 = 66$, we can estimate the answer as about 660.

The $\times$ factor

Suppose you feel a sudden urge to multiply 78×37. This looks nasty. The vital point to remember is that all numbers, with the exception of prime numbers, have smaller factors. For example, $111 = 37 \times 3$, so that $37 = 111/3$. To calculate 78×37, it may be quicker to compute it as $78 \times 37 = 78 \times 111/3 = (78/3) \times 111 = 26 \times 111$, and then use the swift method either for multiplying by 13 and doubling (to multiply by 26), or multiply by 111 swiftly by shunting (see the next section!). Choose whichever factorization you prefer!

One of the glories of pre-decimal currency in England and the Imperial system of measurement may well have to do with factors. Twelve can be divided by 1, 2, 3, 4, and 6, which may be the basis for having 12 inches in a foot and 12 pennies in a (pre-decimal) shilling. Likewise the number 60 can be divided by 1, 2, 3, 4, 5, and 6. (It can be divided exactly by other numbers as well, but we only need to check up to the square root of 60. The other factors will be the complements of these; that is, the numbers you need to multiply by these in order to get 60. The other divisors are 60, 30, 20, 15, 12, and 10.) The number 60 having so many integers as divisors may explain why the Ancient Babylonians used the sexagesimal system, counting in 60 s. We acknowledge this in the metric system, where a right angle remains $90°$. The gradian, with 100 gradians in a right angle, never caught on. Sure, the $45 - 45 - 90°$ triangle becomes a 50–50–100 gradian triangle (angles in a triangle add up to 200 gradians), but the other standard triangle goes from $30 - 60 - 90°$ to $33.333 - 66.666 - 100$ gradians, which is user *un*friendly.

Shunting for show

It's just a jump to the left, and then a step to the right.
"The Time Warp"

To be brief—which helps saves paper and therefore trees—I'll use the word shunting. To shunt, I mean shift a number over to the left or to the right. The number 1, shunted two places to the left, is 100, and two places to the right it becomes 0.01.

A simple example of shunting is if you want to multiply 47×111. Simply write 47 down three times, but shunted over one place each time, to get:

4700

470

47

This you add up swiftly to get 5,217.

How?

There are several books available on the subject of quick calculations— not to mention some websites and YouTube videos. These can be good at teaching the tricks and techniques in carrying out a calculation. Edward H. Julius, for example, has written several books on Rapid Math. A shout out should also go to Jacow Trachtenberg, who developed an entire system for rapid calculations while detained in a Nazi concentration camp.[3]

Few, if any, other books explain *why* or *how* a particular method works. This book does: many sections have short explanations of why things work. These nearly always tie in to some simple formulas from high-school mathematics, ones you may never have used to accelerate your arithmetic.

Try these

Estimate three ways: first overestimate, then underestimate, and finally, construct a "best estimate." Do *not* work out the answer—the idea here is to learn to estimate rapidly. Sometimes, rapid estimates form part of offbeat job interviews. "How many ping pong balls can fit in a 747 airplane" is a standard non-standard question (go ahead, find the data, and come up with your own estimate!) or how many babies are born each year in the UK. (If there are 66 million Brits who live on average to 75, and the population isn't growing, that's about 66/75 million babies, or 880,000 little ones each year.) Here are some far more straightforward "plug and chug" questions to get you started.

1. 4π
2. 1.87×2.34
3. $24,000 \times 26$
4. 0.007×265
5. $0.05/547$
6. $547/0.051$
7. $456 - 4.63$
8. $0.93 - 0.296$
9. $412 + 891$
10. $1,075 + 2,342$

---

[3] For his method, see Jacow Trachtenberg, *Speed System of Basic Mathematics* (New York: Souvenir Press, 1989).

2

Speedier Subtractions and Sums

Life is painting a picture, not doing a sum.

OLIVER WENDELL HOLMES, JR.
"The Class of '61," from *Speeches* (1913)

There is no real making amends in this world, any more nor you
can mend a wrong subtraction by doing your addition right.

GEORGE ELIOT (MARY ANN CROSS [NÉE EVANS]),
Adam Bede, chapter 18

England is known for many things, one of which is its pubs. If you head
to the Rose and Crown one evening, you will probably find the locals
enjoying a game of darts—a sport so popular in Great Britain that it is
frequently televised. Players usually start with a score of 501 and throw
three darts, carrying on until one of them—either by landing on a
double or hitting the bull's eye—ends up at zero.

The impressive thing is that most darts players, if they score 57,
say, can work out their new lower total rapidly. Should you want to
compute more quickly, keeping score in a darts match is a good way to
develop that skill (and, for card players, cribbage also hones your talent
for totting up).

Suppose the game comes down to the last few darts, and the player
has 81 before scoring 57. Traditionally, you would calculate the new
score by writing:

$$81-$$
$$\underline{57}$$

Things turn nasty, rapidly. You can't take 7 from 1, you have to
"borrow" a 10 from the tens column. So, you cross out the 8, replace it
by a 7, then add a 10 to the ones column. This looks like:

$$7\cancel{8}\ ^1 1-$$
$$\underline{5\quad 7}$$

And you now take the 7 from the 11 to get 4, and the 5 from the 7 to get 2. Hence:

$$81-$$
$$\underline{57}$$
$$24$$

This is not exactly swift at all. Luckily, there are ways to speed this up, as the next section shows. (Speaking of darts and mathematics, there are ways to arrange the numbers from 1–20 in such a way as to punish inaccurate throwing and that are more difficult than the standard dartboard.[1])

Subtract by adding

Humans are swifter and surer when adding, rather than subtracting. When faced with $81 - 57$, ask yourself "what must be added to 57 to make 81." This is far quicker than the long and complicated calculation in the section before. The answer is 24.

Similar, but slightly different, is to count off in tens. That is to say, start at 57 and add tens, something we humans are extremely good at. Hence 57, 67, 77. Keep track of the number of tens you have added—in this case, two (fingers work well for this purpose). You are left to compute $81 - 77$, which is quick and easy to do (the answer is 4). Add this to the 20 you've already added to obtain 24.

As a second example, consider $2.73 - 1.42$. Making an estimate, we know this is roughly $2 - 1 = 1$. Erase the decimal points and consider $273 - 142$. Subtract the hundreds, which leaves $173 - 42$. Count in tens as before, to generate 52, 62, 72. We can do $173 - 72 = 100 + (73 - 72) = 101$, which, when we include the 30 that we already added in the "counting by tens" process, gives us 131. As the answer is about 1, we must have $2.73 - 1.42 = 1.31$.

Two-step subtraction

Faced with the same problem, $81 - 57$, there is another approach. It is easier to add or subtract multiples of 10. Hence, you can convert

[1] Trevor Lipscombe and Arturo Sangalli, "The devil's dartboard," *Crux Mathematicorum*, 27(4) (2000), 215–17.

this from a complicated one-step calculation to an easy two-step calculation. After all, we know that $60 = 57 + 3$. Hence $81 - 57 = 81 - (60 - 3) = (81 - 60) + 3 = 24$. While this is cumbersome written down on a page, it is something you can do rapidly in your head: to subtract 57, don't. Instead, subtract 60, then add on 3.[2]

Beauty is in the eye of the beholder. If you prefer, you could think of $81 - 57$ as $80 - 57 + 1$, "80, subtract 57, add 1," which gives $23 + 1 = 24$.

Choose again our decimal problem $2.73 - 1.42$. Remove the decimal points, to make things easier. Then write $273 - 142 = 273 - 140 - 2 = 133 - 2 = 131$. Putting the decimal points back in to get an answer of the right size, $2.73 - 1.42 = 1.31$. Or, you could form $273 - 142 = 273 - 143 + 1 = 130 + 1 = 131$. Put in the decimal point for the answer 1.31.

Double-switch subtraction

Feeling confident? If so, you can subtract swiftly by changing *both* the numbers involved at the same time. That is to say, following our example, you can write $81 - 57 = (80 + 1) - (60 - 3) = (80 - 60) + 1 + 3 = 20 + 1 + 3 = 24$.

Likewise, $273 - 142 = 270 - 140 + 3 - 2 = 130 + 1 = 131$, so that $2.73 - 1.42 = 1.31$.

Be warned: doing two steps at once is both fast and flashy, but it significantly increases your chances of making a mistake. Or, to be more accurate, when I do so, it increases my chances of making a mistake.

Subtraction either side of an "easy" number

Suppose you have a number just over 100 (143, say) and need to subtract a number just under 100 (86, say). To carry out the subtraction, work out by how much the first number exceeds 100, in this example 43, and how much less than 100 the other number is, which here is 14. Add them to get the answer: $143 - 86 = 43 + 14 = 57$. Multiples of 100 are easy, in this sense, as is 1,000.

If we seek $1.23 - 0.96$, say, the same principle applies. Straight away we clean out the decimal points and consider $123 - 96$. Sum the 23 and 4 to get 27. But we know our answer must be about 0.3, so our answer is $1.23 - 0.96 = 0.27$.

---

[2] To be somewhat formal, we are making use of the associative law of arithmetic, that $(a + b) + c = a + (b + c)$, and the commutative law, that $a + b = b + a$.

How it works: we have a number $(100 + n)$, which is n more than 100, and another number $(100 - m)$, which is m less than 100. Subtracting gives $(100 + n) - (100 - m) = 100 + n - 100 + m = n + m$.

Before turning attention to addition, let's take a moment to use our newfound subtraction skills. Consider a three-digit number, any one you like. If you want to book a vacation in Disneyland, California, the area code is 714. Now write its mirror image, so to speak, by swapping the units and the hundreds digits, which is 417. Take the smaller number away from the larger, to get $714 - 417 = 297$. Now form the mirror number again, which is 792, but this time add: $297 + 792 = 1,089$. If you aspire to study mathematics at one of the world's leading universities for mathematics, you might need to call Harvard or MIT, whose area code is 617. The mirror number is 716 and $716 - 617 = 99$, or, as we prefer to write it, 099. Mirror this to get 990 and add, giving $990 + 99 = 1,089$. Amazing!

How it works. We start with $100n+10m+p$, whose mirror is $100p+10m+n$. Subtracting gives $100\left(n - p\right) + \left(p - n\right)$. Employ a cunning plan: add 0. Or, rather, add $-100+90+10$. Adding these in, we get $100\left(n - p - 1\right) + 90 + \left(10 + p - n\right)$. Mirror to get $100\left(10 + p - n\right) + 90 + \left(n - p - 1\right)$. The last step is to add the two, which is $100\left(n - p - 1 + 10 + p - n\right) + 180 + \left(10 + p - n + n - p - 1\right)$. Simplifying, this is $900 + 180 + 9 = 1,089$.

Two-step addition

By the same token, if you are asked to add $79 + 64$, it is quicker to think "$64 = 60 + 4$." Then calculate $79 + 64 = 79 + (60 + 4) = (79 + 60) + 4 = 139 + 4 = 143$, which is quicker than writing the sum down, carrying over numbers from the units to the tens column, the tens column to the hundreds column, and so on. We could also do a double switch, setting $79 + 64 = (80 - 1) + (60 + 4) = 140 + 4 - 1 = 143$.

The calculation $3.76 + 2.89$ goes the same way. As an estimate, the answer is more than $3 + 2 = 5$. Dispense with the decimals and write this as $376 + (290 - 1) = (376 - 1) + 290 = 375 + 290 = 665$. As the answer is more than 5, we know we must have $3.76 + 2.89 = 6.65$. Depending on how you feel, you could add an extra step to make things quicker by writing $375 + 290 = 375 + (300 - 10) = (375 + 300) - 10 = 675 - 10 = 665$, which, as before, gives 6.65 when you reinsert the decimal.

Use whichever approach gives you the correct answer the fastest. Remember, it is long-winded to write down all the intermediate steps to show the way of thinking but, in practice, you can do the sum extremely quickly.

Try these

As a reminder, negative numbers exist. Remember that $-n + (-m)$ is identical to $-(n + m)$. As in, add the two numbers as if they were both positive, then flip the sign. A sum of the form $-n + m$, on the other hand, is the same as $m - n$, so switch the order and proceed as usual.

1. $7.56 + 6.19$
2. $-8,230 - 5,800$
3. $-50.2 + 51.1$
4. $0.96 + 0.801$
5. $-963 - 226$
6. $34.6 + 5.7$
7. $-8.00 - 4.67$
8. $87,100 + 40,900$
9. $-82.8 + 3.20$
10. $38.2 + 71.9$
11. $98.3 + 11.7$
12. $-0.57 + 4.13$
13. $-68.80 - 8.10$
14. $-24.90 - 9.23$
15. $-9.50 - 9.70$
16. $-121 - 803$
17. $41.6 - 44.3$
18. $0.382 + 0.417$
19. $8.27 + 8.23$
20. $57.7 - 69.6$
21. $-8,130 - 4,750$
22. $-3.48 - 1.50$
23. $516 - 461$
24. $-96.2 + 86.1$
25. $50.80 + 0.76$
26. $5.04 - 2.26$
27. $-27.3 - 11.5$
28. $4,980 - 3,690$

29. $72.9 - 50.4$
30. $2.5 - 94.5$
31. $-82.4 - 51.9$
32. $32.8 + 31.7$
33. $-937 + 272$
34. $88.4 + 59.1$
35. $-9.48 + 3.48$
36. $48.5 - 24.8$
37. $36.0 + 30.5$
38. $6.12 - 6.90$
39. $-5.17 + 9.35$
40. $83.9 - 34.3$
41. $1.54 - 6.27$
42. $24.7 - 85.7$
43. $-9.31 + 5.92$
44. $3,370 + 6,890$
45. $20.6 - 56.6$
46. $2.05 + 6.68$
47. $-4,834 + 231$
48. $5.22 - 3.48$
49. $79.8 - 35.2$
50. $376 - 482$

Interlude I

The Magic of 111,111

Today is my one hundred and eleventh birthday: I am eleventy-one today!

BILBO BAGGINS, in J.R.R. Tolkien's *Lord of the Rings*

With a new-found ability to subtract at speed, you can impress friends with a "magic" trick that, as it's based on math, never fails (this is but one example of "mathemagic," about which many books have been written[1]). Write down on a sheet of paper four numbers that add up to 111,111. For effect, you can write these on separate pieces of paper and keep them in a small pouch worn around your neck. Write down a fifth number—a dummy number—that forms the basis for the trick. It can be any number you like, for you won't actually use it! But you *do* need to know it's the dummy number and not one of the four that add up to 111,111. Next, ask your friend, or the class, or an audience member to write down a six-digit number. You show them how to do this (no help is required, but this is magic, after all) by writing down the first digit for them. Just to show them, just to help them, just to get them started—whatever patter you prefer. This leading digit must be a 1. Tell them to write down the next digit, "a number between 1 and 9." It's highly unlikely that they will choose 1, given that you've just written that number down. Eventually, they will have generated a six-digit number, none of which contains a zero. Suppose this is 127,975. With a great flourish, produce the pieces of paper prepared earlier and ask the person to write down the four numbers down, one at a time, as you read them out. Don't show your friend any of the pieces of paper, and don't show them the dummy number. Instead, hold that piece of paper up as though reading it, the same as for all

---

[1] Persi Diaconis and Ronald Graham, *Magical Mathematics: The Mathematical Ideas that Animate Great Magic Tricks* (Princeton, NJ: Princeton University Press, 2015) is a famous book on the subject.

the others, and ask them to write down the number 16,864. Why? Well, that's the number you get when you subtract one from each of the digits in 127,975. The person then has to add up all five numbers—you may need to give them a calculator—and, lo and behold, they have exactly the number they wrote down to begin with. This works because $127,975 = (127,975 - 111,111) + 111,111$. Your four numbers add up to the 111,111, whereas $127,975 - 111,111 = 16,864$. As long as no-one sees your dummy number—which you replaced by 16,864 in this example—they'll be amazed. Or at least mildly appreciative.

There are other similar math tricks, such as one proposed by Walter Sperling,[2] in which the mysterious Herr Pfiffig[3] has a trick. Write down three numbers, he supplies three of his own, and—miraculously—the six of them add up to 2,997. If we give him 237, 489, and 632, Pfiffig writes down 367, 510, 762, and wunderbar, these six numbers sum to 2,997.

There is no real trick here, if you spot that $2,997 = 3 \times 999$ and note that $999 - 237 = 762$, $999 - 489 = 510$, and $999 - 367 = 632$. By using 999, you won't have to worry about having to borrow numbers from the next column, and reversing the order of the numbers you are given hides what you're doing, at least a little bit, adding to the wow factor. And last, 2,997 is not a nice round number, so your feat seems extra-prodigious to a non-mathematical audience!

To return to 111,111, and to go beyond addition, it's a great number to square. That's because $111,111^2 = 12,345,678,987,654,321$, which also makes it a mathematical palindrome, a number that is the same when written backward as it is forward.

---

[2] Walter Sperling, *Auf du und du mit Zahlen* (Zürich, Switzerland: Rüschlikon, 1955), p. 75.

[3] Pfiffig is a German word that conveys the idea of someone who has it all: intelligence, competence, together with good looks, charm, and allure.

3

Accounting for Taste: Calculating Columns Quickly

They must feel the thrill of totting up a balanced book
A thousand ciphers neatly in a row
> From "A British Bank (The Life I lead),"
> RICHARD AND ROBERT SHERMAN

Parliament has given us the powers to put down Fifth Column activities with a strong hand.
> WINSTON CHURCHILL, addressing Parliament, June 4, 1940

There are long lists of numbers waiting to be summed, even in the unlikeliest of places. Genesis, the first book of the Hebrew Bible, for example, lists the ages at which the patriarchs died. From this, people have computed the years that must have passed since the foundation of the world. Biblical chronologists include religious adherents: Rabbi Yose Ben Halafta, in the second century, came up with an estimate of 3761 BCE, while the seventh-century English polymath, Saint Bede the Venerable, computed 3952 BCE. Scientists also came up with answers, such as Johannes Kepler (3992 BCE), famous for his three laws of planetary motion. Most famously, Archbishop James Ussher (1581–1656), of the Church of Ireland, proposed 4004 BCE and, more specifically, "This beginning of time, according to our chronology, happened at the start of the evening preceding the 23rd day of October."[1] These results

---

[1] James Ussher, *Annals of the Old Testament, deduced from the first origins of the world, the chronicle of Asiatic and Egyptian matters together produced from the beginning of historical time up to the beginnings of Maccabees* (1658). While the universe might not have actually been born on the evening of October 22, Satan was. Or, more formally, Miroslav Šatan, a Slovakian ice hockey player whose teams have included Canada's Edmonton Oilers and the United States' New York Islanders, who was born on October 22, 1974.

conflict with the observations of modern cosmologists.[2] If Ussher is right, and that's a big if, those alive on October 22, 1996 witnessed the universe's 6,000th birthday.

Saint Francis of Assisi is one of the most beloved figures in medieval Europe. He had a great love for nature and, furthermore, described himself as being married to "Lady Poverty." It is ironic that a later Franciscan and mathematician, Luca Pacioli (ca. 1447–1517), became the father of accounting. Although the system of double-entry book-keeping had been known for some time (since the early medieval period), Pacioli spread knowledge of it widely through his book *Summa de arithmetica, geometria, proportioni et proportionalita*, published in 1494. To do accounting properly, though, you need to add up long columns of numbers—accurately and rapidly. Whether you wish to check your finances or to calculate your own Bible chronology, there are ways to do so.

But there are non-accounting reasons to add columns of numbers quickly, as it allows you to check for mathematical patterns. If you have a hunch something mathematical is true, you can check your intuition rapidly. As an example, take three different digits, such as 3, 1, 4—the first three digits of π—and form every possible three-digit number from them. Being methodical, list them from smallest to largest as 134, 143, 314, 341, 413, 431. Now add, as swiftly as you can. The result, pleasing to Americans, is 1,776. Let's take another example, $9^3 = 729$. We can form 279, 297, 729, 792, 927, and 972. These total 3,996. Switch on the mathematical pattern-spotting part of the brain. Can we see any connection between the two numbers? Well, the numbers 1, 3, 4 add up to 8, and $8 \times 222 = 1,776$. On the other hand, 2, 7, and 9 sum to 18, and $18 \times 222 = 3,996$. By being able to add (and divide) rapidly, we've stumbled on something interesting about playing with three-digit numbers, that if the three digits sum to n, the sum of the various permutations of those digits sums to $222n$.[3]

---

[2] An intriguing book by mathematician Florin Diaçu tells the story of chronologies, including a group who believe that human history is 1,000 years shorter than we think. *The Lost Millennium: History's Timetables Under Siege* (Baltimore, MD: Johns Hopkins, 2011).

[3] If the digits are n, m, p, we know that n will be in the hundreds column twice, the tens column twice, and the units column twice, which sum to $222n$. The same is true of m and p, so the sum is $222\left(n + m + p\right)$.

Adding numbers "mystically"

Given a long string of one-digit numbers to add up, stare at them (hence "mystically") and look for those that add up to 10. To make things less opaque, an example may help.

The puritans left England on the Mayflower to seek religious freedom and arrived close to Cape Cod on November 11, 1620. Eventually, 13 colonies came into existence before the American War of Independence, or the Revolutionary War. The Colonists displaced native Americans, but some tribes still exist in the former colonies and are federally recognized. These are:

Colony	Native American Tribes[4]	Total
	(Federal + State recognized)	
Massachusetts	2 + 1	3
New Hampshire	0 + 0	0
Connecticut	2 + 3	5
Rhode Island	1 + 0	1
Delaware	0 + 2	2
New York	8 + 3	11
New Jersey	0 + 3	3
Pennsylvania	0 + 0	0
Virginia	7 + 11	18
Maryland	0 + 2	2
North Carolina	1 + 7	8
South Carolina	1 + 7	8
Georgia	0 + 3	3

To add up the number of Native American tribes in the 13 colonies, we seek the sum $3 + 5 + 1 + 2 + 11 + 3 + 18 + 2 + 8 + 8 + 3$.

Insert brackets to set off numbers that combine to form some multiple of 10. Clearly, there are plenty of options for how to combine these numbers, but experience suggests you want to pair up the large numbers if possible, as the tough part is summing the numbers left over at the end; the smaller these are, the easier that task will be. We can introduce brackets and reorder the sum to get:

---

[4] NCSL (2020) "Federal and state recognized tribes." https://www.ncsl.org/research/ state-tribal-institute/list-of-federal-and-state-recognized-tribes.aspx#State (accessed August 1, 2020).

$$3 + 5 + 1 + 2 + 11 + 3 + 18 + 2 + 8 + 8 + 3 =$$
$$3 + 5 + [1 + 8 + 11] + 3 + (18 + 2) + \{2 + 8\} + 3$$

Which we now shuffle into an order that is easier to add:

$$3 + 3 + 3 + 5 + 20 + 20 + 10 = 64.$$

A piece of advice. Notice how each clump of numbers that adds up to 10 has its own distinctive "envelope," whether parentheses (. . .), brackets [. . .], or braces { . . . }, which, if necessary, are repeated. So, next up, you could use ((. . .)), [[. . .]] if you had more than three clumps. This helps you keep track of the sum and keep down the errors (or trace them) far more easily.

Cricket scores present a greater challenge. Recently, the Vatican assembled a cricket team. On a tour of England, they combined forces with one from the Church of England to compete against a "Multifaith XI." The match, somewhat ironically held at Lord's, was reported by the *Church Times* on July 13, 2018.[5] The Vatican/Canterbury XI wielded the willow first and their batsmen drove in the following number of runs:

Wright	2
Rylands	31
Kennedy	34
Ferdinado	6
Lee	34
Markose	16
Watkins	31
Paulson	8
Marshal	11
Matthew	0
Ettolil	2

To determine the number of runs scored by the team, we seek to add:

$$2 + 31 + 34 + 6 + 34 + 16 + 31 + 8 + 11 + 2$$

[5] Stephen Fay (2018, July 23) "Cricket: Vatican and Canterbury Joint XI coast to victory against multifaith side." *Church Times*. https://www.churchtimes.co.uk/articles/ 2018/13-july/news/uk/cricket-vatican-and-canterbury-joint-xi-coast-to-victory-against -multifaith-side (accessed October 30, 2020).

By our process of mystical inspection, we recast this as:

$$2 + 31 + (34 + 6) + [34 + 16] + \{31 + 8 + 11\} + 2$$
$$= 33 + 40 + 50 + 50 + 2 = 175$$

This is curious, as with the 14 extras reported for no balls, wides, and the like, the total should be 189. The *Church Times*, instead, printed in their scorecard that the Vatican/Canterbury XI scored 185 runs, and reiterated it in the main article, saying "The multifaith team were set 186 to win." They could not do so, amassing only 124 runs, but at least the *Church Times* got the Multifaith XI's score correct! Perhaps the number of extras scored by the Vatican/Canterbury side were only 10, rather than 14. After all, scorekeepers are not infallible.

Adding columns without carrying

When adding long lists of numbers, "carrying over" burdens the person doing the sum. But there's no need to. Going back to a point made earlier on—both in this book and one published in 1543—by keeping your digits neatly aligned in a column, you can greatly boost your column-adding speed. It turns out there are a variety of methods to choose from. Try them all and see which you feel most comfortable with.

Tallying in tens

The verb "to tally" has a long history, and tally sticks—simple devices with which to count, or to record debts—date back to Paleolithic times. If you borrowed three units of money from me, say, we could take a stick and record three notches in it. We split the stick down the middle— you keep your half with its three notches and I keep my half. That way, we agree on the amount of cash I gave to you. Different thickness of notches could stand for different denominations. According to *The Dialogue Concerning the Exchequer*, a twelfth-century manuscript of Richard FitzNeal, the sum of 1,000 pounds was recorded by a notch the width of your palm; 100 pounds by the width of a thumb (about an inch); 20 pounds by the breadth of a little finger; and a pound by a mark "the width of a swollen barleycorn." There was just one problem: the sticks can't be identical, otherwise you could claim I owed you the money. To avoid this, we lop a piece off the end of your stick. Whoever had the

shorter stick owed the money. I was a stick holder, which evolved into
the modern word stockholder. Tally, as a word in itself, comes from
a word for twig, showing that cutting notches into wood was an old
custom. Indeed, the Ishango bone has similar notches in it; found in
the Democratic Republic of Congo back in the 1960s, it dates back to
between 18,000 and 20,000 BCE.

One way to tally a column of figures swiftly, without the aid of a stick,
is simply to count in tens. An example shows the method. Charlton
Athletic, a football club based in South East London, spent several years
in the Premier League. The goals they scored during those years were as
follows, together with the thought process of "counting in tens" to add
them up swiftly.

Goals scored	Thought process
41	41
50	91
38	101, 111, 121, add 8: 129
45	139, 149, 159, 169, add 5: 174
51	184, 194, 204, 214, 224 add 1: 225
42	235, 245, 255, 265, add 2: 267
41	277, 287, 297, 307 add 1: 308
34	318, 328, 338, add 4: 342

Their goal tally for their eight seasons in the Premiership is 342, which
goes a long way to explaining why they are no longer in the top flight
of English football. Over the same period in which they scored 342,
Charlton let in 442, outscored by a stunning 100 goals. As a lifelong fan
of Charlton Athletic, though, I am used to disappointment.

Turning to happier things, recall the popular Christmas carol, "The
Twelve Days of Christmas." In it, the singer reports receiving a large
number of gifts from their true love, who seems to have a great interest
in birds. On the first day, the singer receives a partridge (in a pear tree).
Day two brings another partridge, but also two turtle doves. Day three
heralds the arrival of three French hens. So it continues throughout
the 12 days stretching from Christmas to the Epiphany, the feast that
marks the arrival of the gift-giving Magi. We turn our calculation skills
to working out how many Christmas presents the singer receives.

There is one partridge received on 12 days, for a total of 12 partridges.

Two turtle doves show up each day for the next 11 days, for a total of
22 turtle doves.

Three French hens are unwrapped on the next 10 days, for a total of
30 French hens.

All in all, then, we need to sum:

- $1 \times 12 = 12$
- $2 \times 11 = 22$
- $3 \times 10 = 30$
- $4 \times 9 = 36$
- $5 \times 8 = 40$
- $6 \times 7 = 42$

Thanks to the power of symmetry, we can double this to get the total number of gifts. After all, on the last day the rather noisy twelve drummers drumming arrive, but for one day only. Thus 12 gifts for 1 day, as opposed to the one partridge-laden pear tree each day for 12 days.

Number	Thought process
12	
22	12, 22, 32, add 2: 34
30	64
36	74, 84, 94 add 6: 100
40	140
42	180, add 2: 182

The true love of the singer sent along a total of 364 gifts from Christmas to Twelfth Night. The first 100 gifts were birds, supplemented later on by 42 swans (6 lots of 7 swans a swimmin') and 42 geese (7 lots of 6 geese a layin') for a grand total of 184 avian lifeforms. The song does not tell us how the recipient of these gifts reacted, or whether all of the gifts are entirely legal.[6]

Circular sums

The plan here is to go *down* the column of numbers adding the tens, and then go *up* the column adding the units. Again, the method is best illustrated by an example, which is the age in years, at election,[7] of the first 10 American presidents.

[6] In England, almost all swans are owned by the Crown, with the monarch serving as "Seigneur (Lord) of the Swans." On the River Thames, the swans belong jointly to the Crown, the Guild of Vintners, and the Guild of Dyers. So how the gift giver gets the swans is a mystery.

[7] Age at election is slightly different from age at which they take office. As some presidents died in office, their successors would have assumed office rather than being elected to it.

Name	Age
George Washington	57
John Adams	61
Thomas Jefferson	57
James Madison	57
James Monroe	58
John Quincy Adams	57
Andrew Jackson	61
Martin van Buren	54
William Henry Harrison	68
John Tyler	51

The thought process—your calculation in progress—should go something like this: 50, 110, 160, 210, 260, 310, 370, 420, 480, 530. That is to say, you've added the tens column from top to bottom. Now, to save time, you go from bottom to top. Starting with 530, you now think 531, 539, 543, 544, 551, 559, 566, 573, 574, 581. The average age of the first 10 presidents, then, was 58.1.

There is a piece of advice here that is worth contemplating. There isn't much variation in ages. If you subtract 50 from the ages of each of the 10 presidents, you get: 7, 11, 7, 7, 8, 7, 11, 4, 18, and 1, which is far simpler to add. We have 500 years so far (50×10) to which we can add 28 (4×7) to get 528 and then add 22 (2×11) to get 550. The total ages then are $550 + 8 + 4 + 18 + 1 = 581$.

As a second example, consider the ages at election of John F. Kennedy (the second youngest president after Teddy Roosevelt) and the 11 presidents who followed him, which includes the oldest one elected so far, Joe Biden.

Name	Age at becoming president
John F. Kennedy	43
Lyndon B. Johnson	55
Richard Nixon	56
Gerald Ford	61
Jimmy Carter	52
Ronald Reagan	69
George H. W. Bush	64
Bill Clinton	46
George W. Bush	54
Barack Obama	47
Donald Trump	70
Joe Biden	78

The addition process, going down and adding the tens column, is: 40, 90, 140, 200, 250, 310, 370, 410, 460, 500, 570, 640. Then, adding the units column from the bottom up: 648, 648, 655, 659, 665, 669, 678, 680, 681, 687, 692, 695. The average age of these presidents is 57 years, 11 months, which is a tad younger than the average age for the first 10 presidents (58.1 years). The second batch of presidential ages, however, displays far wider variation, with three presidents in their 40s, three in their 60s, and two in their 70s. (To be speedy in working out the average age, note there are 12 presidents in the list and there happen to be 12 months in the year. So, whatever the remainder is when you divide 695 by 12, that must be the months you need to tack on.)

The method here can be generalized. What's happening is that we add the units and the tens separately, then join them together. You can do this with all columns in a sum.

Legendary Chicago gangster Alphonse "Al" Capone eventually went to prison, where he died, for the crime of tax evasion. The Inland Revenue Service argued that, even though his only income was illegal, he still had to pay tax on it. As our example, look at the figures for the tax Capone owed:

Year	Tax owed (to the nearest US dollar)
1924	32,489
1925	55,379
1926	39,963
1927	45,558
1928	30,054
1929	15,819

Add up the units column, to get: 42
Add up the tens to get: 32
Add up the hundreds to get: 29
Add up the thousands to obtain: 26
Add up the 10K column to yield: 19

Now write down the simpler sum:

$$190,000$$
$$26,000$$
$$2,900$$
$$320$$
$$42$$

This is straightforward to tally, giving Capone's unpaid tax bill as $219,262. As the cliché goes, that was a lot of money in those days, which is in part why he was sentenced, on October 17, 1931, to 11 years in prison.

Cancel as you calculate

To add a column of single-digit numbers quickly, keep track of the units and cancel out a number by crossing it out once you have reached 10. At the end, you'll have a single digit in your head, and the number of tens you need to add equals the number of digits you've crossed out. Again, this may sound obscure, but an example will help show the method.

Or, rather, an obscure example shows the method. For a reason that (I hope) becomes clear, list 20 English towns, in alphabetical order, as well as their population. The obscure thing is that we use just the first digit of their population. A town with 10,000 people, a city with 100,000, or a conurbation of 1,000,000 therefore are all listed as a "1."

City	Lead digit	Thoughts
Birmingham	1	1
Blackpool	1	2
Bolton	1	3
Bradford	2	5
Brighton	1	6
Bristol	3	9
Coventry	2	11: cross out the 2, remember 1
Derby	2	3
Dudley	1	4
Huddersfield	1	5
Hull	2	7
Ipswich	1	8
Leeds	4	12: cross out the 4, remember 2
Leicester	2	4
Liverpool	4	8
London	6	14: cross out the 6, remember 4
Luton	1	5
Manchester	4	9
Middlesborough	1	10: cross out the 1, remember 0
Milton Keynes	1	1

The last number we have to remember is 1, and we have crossed out four numbers, so the total is 41.

As another example, let's do the same with 20 American cities. Again, these are posted in alphabetical order.

City	Lead digit	Thoughts
Abilene, TX	1	1
Akron, OH	1	2
Albuquerque, NM	5	7
Alexandria, VA	1	8
Allen, TX	1	9
Allentown, PA	~~4~~	10: cross out the 1, remember 0
Amarillo, TX	1	1
Anaheim, CA	3	4
Anchorage, AK	2	6
Anna Arbor, MI	1	7
Antioch, CA	1	8
Arlington, TX	~~3~~	11: cross out the 3, remember 1
Arvada, CO	1	2
Athens, GA	1	3
Atlanta, GA	4	7
Augusta, GA	1	8
Aurora, CO	~~3~~	11: cross out the 3, remember 1
Aurora, IL	2	3
Austin, TX	~~9~~	12: cross out the 9, remember 2
Bakersfield, CA	3	5

The last unit we have in our thoughts is a 5, and there are 4 numbers crossed out, so the total is 45.

The method works extremely well, and quickly, for long strings of single-digit numbers.

The answers are intriguing. The English cities list totaled 41 and the American cities list totaled 45. "Common sense" suggests that every digit should be random, between 1 and 9. With 20 cities on each list, then, we might reasonably expect an answer closer to 90, rather than 40. The obvious rebuttal is that we've only looked at 20 cities, so if we looked at a larger, statistically representative sample, surely we would get an average answer of about 5. We won't.

Simon Newcomb, an astronomer from Nova Scotia in Canada, became one of the first professors of mathematics at the newly founded, and newly funded, Johns Hopkins University in Baltimore (some people have suggested that he is the inspiration for Sherlock Holmes's nemesis, the mathematics professor James Moriarty). Newcomb noticed something strange, inspired by a book of logarithms. The first few pages were intensely well-thumbed, as though logarithms beginning with smaller digits were more common. From this observation, he published an article "Note on the frequency of use of the different digits in natural numbers" that gave an explanation.[8] Unfortunately, it was overlooked. It wasn't until a physicist, Frank Benford, explored the subject in a 1938 paper "The Law of Anomalous Numbers" that Benford's Law became widely known.[9,10] Benford showed that small digits occur often in such things as populations (as we've just seen) and the street addresses of famous people, and the numbers printed in newspaper and magazine articles aren't distributed at random from 1–9, but are far more likely to be 1s and 2s[11] (about 30% will be the digit 1, 17% the digit 2, 12.5% the digit 3).[12]

The Newcomb–Benford Law even finds use in fraud detection. While you can make up the numbers easily, it's hard to fake statistics. Forensic accountants use the Newcomb–Benford law to test whether a company is cooking the books. After all, if you are making up fraudulent receipts for something you didn't actually buy, will you remember to make sure that the cash values obey the Newcomb–Benford Law?[13] Following the election of Joe Biden in the United States, there was some debate

[8] Simon Newcomb, "Note on the frequency of use of the different digits in natural numbers," *American Journal of Mathematics*, 4(1) (1881), 39–40.

[9] Frank Benford, "The law of anomalous numbers," *Proceedings of the American Philosophical Society*, 78(4) (1938), 551–72.

[10] There's another law, Stigler's Law of Eponymy, which says that laws aren't usually named after the people who first came up with them. Stigler wrote this in 1980. Fittingly, Stigler's idea had been proposed earlier, in 1972, by Hubert Kennedy, who dubbed it "Boyer's Law." Fans of Simon Newcomb would say that Benford's Law is a case in point.

[11] The chance of having a first digit d is $P(d) = \log_{10}\left(1 + \frac{1}{d}\right)$. The probability for having the second digit s is $P(s) = \log_{10}\left(1 + \frac{1}{1s}\right) + \log_{10}\left(1 + \frac{1}{2s}\right) + \log_{10}\left(1 + \frac{1}{3s}\right) + \ldots$

[12] Those who enjoy physics might like the account provided in Don S. Lemons, "On the numbers of things and the distribution of small digits," *American Journal of Physics*, 54(9), (1986), 816–17.

[13] See J. Carlton Collins, "Using Excel and Benford's law to detect fraud," *Journal of Accountancy*, April 2017. https://www.journalofaccountancy.com/issues/2017/apr/excel-and-benfords-law-to-detect-fraud.html (accessed November 5, 2018).

over whether Benford's Law could help detect voter fraud in elections. Scholars believe using Benford's Law could be problematic.[14] Mind you, there's nothing special about counting in tens. It's often best to think and make a decision ahead of time before beginning to count. A case in point is with historical data. Henry Mayhew published a book in 1851 called *London Labour and the London Poor* in which a scavenger (someone we would now call a street sweeper) reported his expenses for the week. These were, in terms of shillings (s) and pennies (d) as follows:

	s	d
Rent	1	9
~~Washing~~		3
Shaving		1
Tobacco		7
~~Beer~~	2	4
Gin	1	2
~~Cocoa~~		10 ½
Bread	3	6
Boiled beef	2	4
Pickles and onions		1 ¾
~~Butter~~		1
Soap		1

When the book was published, in 1851, the Industrial Revolution was at high tide, and Queen Victoria had opened the Great Exhibition (of the Works of Industry of All Nations) at the Crystal Palace, which had been constructed in Hyde Park (and subsequently moved to South East London). But English currency still had the half penny, or ha'penny, and the quarter penny, or farthing.[15] What's more, there were 12 pennies in 1 shilling (and 20 shillings in 1 pound). To work out the scavenger's living costs for the week, then, we can use the cross-out method, but cross out pennies only when they sum to 12, not 10.

[14] Joseph Deckert, Mikhail Myagkov, and Peter C. Ordeshook, "Benford's Law and the detection of election fraud," *Political Analysis*, 19 (2011), 245–68.
[15] The farthing, if you see one, is cute, with a British robin on the back. The ha'penny was the basis of the great game of shove ha'penny, a game so old it used to be called shoffe-grote, using a groat, a coin that went out of circulation in 1662.

Going down the shillings column, we add $1+2+1+3+2 = 9$. Going down the pennies column, we add $9 + 3$ to get 0 (cross out the word "washing"). Then add $1 + 7 + 4$, which also equals 12, so cross out the word "beer." Add the 2 and 10 to get 12, and cross out the word "cocoa." Add the $6 + 4 + 1 + 1$ to get 12 and cross out "butter." This leaves the 1 penny for soap, the ha'penny for cocoa, and the 3 farthings for pickles and onions, which sum to 2 ¼ pennies. As we've crossed out washing, beer, cocoa, and butter, that means we've got an extra 4 shillings to add to the 9 we already counted, to make a scavenger's weekly expenses of 13s. 2 1/4 d.

The pricing pandemic

Marketing experts clearly believe there's something psychologically magical about paying just under a certain amount. Cars might be $19,999, but we think we paid just over $10,000 for it, rather than $20,000. A book might be $24.95, which we think of as about $20, rather than $25. Even if there's a bit more honesty and a book is priced at, say, $34.50, sales tax might bump that up to $36.57. When you go to buy a large number of things, there's a simple way to keep track of what's going on. In one column, you add the rounded up value of the price, and in the second column, you note how much you rounded it up.

A quick trip to the local supermarket during the Covid crisis resulted in a bill that looked like this (in dollars and cents):

Donuts:	2.99
Ricotta cheese	2.69
American cheese:	4.97
Frozen waffles	2.99
Pizza	6.49
Breakfast drink	7.59

Apart from concluding that the eating habits portrayed here aren't exactly healthy, we can do something else. Rewrite this in the form of the rounded-up value, and the boost column, the amount of money we had to add to "boost" the actual price up to the rounded-up value. The former is in dollars, the latter in cents.

Donuts:	3	1
Ricotta cheese	3	31
American cheese:	5	3
Frozen waffles	3	1
Pizza	7	51
Breakfast drink	8	41

A quick glance down the left column gives the round sum of $29.00. The right-hand column totals, by whichever method you prefer, $1.28. Hence the amount paid was $29 − $1.28 = $27.72. Perhaps I ought to add up the nutritional values as well...

Estimating columns of numbers

Sometimes, you don't need an exact answer to a calculation, merely a rough estimate. Heralds, for example, knew the coats of arms of knights and, more importantly, how many men that knight would have brought to the battlefield. By identifying the coats of arms at a distance, then, the herald could determine whether his master was out-numbered and should withdraw from the battlefield, or attack at once to press their numerical advantage. Those knights arriving fashionably late at the battle could sum up the size of the two opposing armies and make a last-gasp tactical decision on which noble to support.

One example along these lines comes from the Book of Numbers. There, in Numbers 1–4, we are told the number of warriors from each of the 12 tribes of Israel.

Tribe	Number of warriors
Reuben	46,500
Simeon	59,300
Gad	45,650
Judah	74,600
Issachar	54,400
Zebulon	57,400
Ephraim	40,500
Manasseh	32,200
Benjamin	35,400
Dan	62,700
Asher	41,500
Naphtali	53,400

The simple trick to estimating is this: don't sweat the small stuff. For example, take the first column only, the 10K column, and add. Doing so quickly gives 55. Now add a 6, to get 61. The total, then, is about 610,000 warriors. Why add the 6? We're playing a numbers game and figuring that in the next column, the thousands column, for every number above 5 (which would be rounded up to 10) there's a number below 5 (which would be rounded down to 0). The average is thus 5, and as we have 12 tribes of Israel, that gives 60, which is a contribution of 6 to the 10K column. The actual answer is 603,550, so our quick estimate of 610,000 is not too bad—accurate to within 1.1%.

One of the biggest parties ever was the Field of the Cloth of Gold, a meeting between England's Henry VIII and King Francis I of France. Attendees gorged themselves over 17 days from June 7 to June 24, 1520. A rich historical source, the Letters and Papers, Foreign and Domestic, of King Henry VIII, lists those who accompanied Henry.

> For the King: The cardinal of York, with 300 servants...; one archbishop with 70 servants...; 2 dukes, each with 70 servants. 1 marquis with 56 servants. 10 earls, each with 42 servants... 5 bishops, of whom the bishop of Winchester shall have 56 servants—each of the others, 44 servants... 20 barons, each to have 22 servants... 4 knights of the order of St. George, each to have 22 servants. 70 knights... each knight to have 12 servants. Councillors of the long robe; viz., the King's secretary, the vice-chancellor, the dean of the Chapel, and the almoner, each to have 12 servants... 12 King's chaplains, each with 6 servants... 12 serjeants-at-arms, each with 1 servant... 200 of the King's guard... 70 grooms of the chamber, with 150 servants... among them; 266 officers of the house, with 216 servants...; 205 grooms of the stable and of the armories.... The earl of Essex, being earl marshal, shall have, beside the number above stated, 130 servants.

How many people, approximately, came from England to the Field of the Cloth of Gold? Writing things out, we have:

Cardinal of York[16] with 300 servants	301
One archbishop with 70 servants	71
Two dukes each with 70 servants	142
One marquis with 56 servants	57
Ten earls, each with 42 servants	430
Bishop of Winchester with 56 servants	57
Four unnamed bishops with 44 servants	180
Twenty barons with 22 servants	460
Four knights of Saint George with 22 servants	92
Seventy knights with 12 servants	910
Four Councillors of the long robe with 12 servants	52
Twelve King's chaplains with 6 servants	84
Twelve serjeants-at-arms with a servant	24
King's guard	200
Seventy grooms of the chamber with 150 servants	220
Two hundred and sixty-six officers of the house with 216 servants	482
Grooms of the stable and armories	205
Earl of Essex's extra servants	130

To estimate the size of the retinue, begin by adding up the hundreds only. That gives a sum of 33 or, more correctly, 3,300. As there are 18 entries, we'd add a further 900 for an estimate of about 4,200. According to the *Letters* the "Sum total of the King's company, 3,997 persons." Add up the columns using any of the methods so far to check if this is correct![17]

Perhaps the award for the best quicker-calculation method for addition should go to Carl Friedrich Gauss. The story, possibly even true, is that in school young Gauss was asked to sum the numbers from 1 to 100, the aim being to keep him busy for a while. It didn't. All Gauss did, the legend runs, was to write:

$$1 + 2 + 3 + 4 + \ldots 100 \text{ and underneath wrote}$$
$$100 + 99 + 98 + 97 + \ldots 1$$

[16] This is Cardinal Wolsey. Another attendee was the future Saint Thomas More. The archbishop was the Archbishop of Canterbury, William Warham. The Bishop of Winchester, probably the richest bishopric in England at that time, was Richard Foxe, founder of Corpus Christi College, Oxford.

[17] The Rutland Papers claim that the retinue was some 4,544 men, not included the wives of kings, earls, marquises, and so forth.

Adding them up gives twice the answer, which is 100 lots of 101, or 10,100. The sum of the first 100 numbers, then, is 5,050. If true, it was a masterful stroke of mathematical thinking.[18]

Try these

1. In the United States before the Civil Rights Act of 1964, when segregation remained legal, a number of colleges and universities were established for African Americans. Here are the largest 10 historically black colleges and universities (HBCUs) in terms of enrollment.[19] Combined, how many students do they have?

HBCU	Enrollment
North Carolina A&T State	11,877
Texas Southern	10,237
Florida A&M	9,913
Howard (Washington, DC)	9,392
Prairie View A&M (Texas)	9,219
Jackson State (Mississippi)	8,558
Tennessee State	8,177
North Carolina Central	8,097
Morgan State (Maryland)	7,747
Albany State (Georgia)	6,615

2. Two of the largest HBCUs are in Texas. What's medical education like in the Lone Star State? To get into medical school in America, you need to pass the Medical College Admissions Test (MCAT). Here are the average MCAT scores of the 10 medical schools in Texas. Estimate, total, and compute the average Texas medical student MCAT score.

[18] For those who haven't seen it done, here's how you generalize Gauss's idea. To sum $1 + 2 + \ldots N$, reverse it to get $N + (N-1) + (N-2) + \ldots 1$. Then add. You have $N+1, (N-1)+2, (N-2)+3, \ldots, 1+N$. In total, there are $N/2$ lots of $N+1$, so the sum is $N(N+1)/2$.
[19] HBCU Lifestyle (n.d.) "Largest HBCU in the nation: Top 10 black colleges by enrollment." https://hbculifestyle.com/largest-hbcu-by-enrollment/ (accessed November 3, 2020).

Baylor College of Medicine	515
Texas A&M College of Medicine	510
Texas Tech University (El Paso) Paul L. Foster School of Medicine	507
Texas Tech University, (Lubbock) School of Medicine	508
University of Texas, Austin, Dell Medical School	512
University of Texas, Medical Branch	508
University of Texas Health Science Center, McGovern Medical School	511
University of Texas, Rio Grande	504
University of Texas, San Antonio	509
University of Texas, Southwest	514

3. Homer's *Iliad* contains, in book 2, the so-called "catalog of ships." This lists the ships captained by Achaean warriors who set sail for Troy. How many ships did the Achaeans muster?

Ethnicity	Number of ships	Captain
Athenians	50	Menetheus
Salamineans	12	Telamonian Ajax
Argives	80	Diomedes
Mycenaeans	100	King Agamemnon
Lacedaemonians	60	Menelaus
?	90	Nestor
Arcadians	60	Agapenor
Epeas of Elis	40	Amphimachus
Men of Dulichium	40	Meges
Cephallenians	12	Odysseus
Aetolians	40	Thoas
Cretans	80	Idomeneus
Rhodians	9	Tlepolemus
Symians	3	Nireus
?	30	Pheidippus
Pelasgians etc.	50	Achilles
?	40	Protesilaus
?	11	Eumelus
?	7	Philoctetes
?	30	Sons of Asclepius
?	40	Eurypylus
Lapiths	40	Polypoetes
Eniens, Peraebi	22	Guneus
Magnetes	40	Prothou

4. If you relax by watching soccer rather than reading the Greek classics, you might prefer to approximate, and then sum up, the attendance for home games of the United States women's national soccer team (USWNT) in 2019. Some members of the USWNT brought a legal action, as their pay was less than that of the men's team, who are far less successful in international competition. Here are the data[20]:

Date	Opponent	Attendance
February 27	Japan	14,555
March 2	England	22,125
March 5	Brazil	14,009
April 4	Australia	17,264
April 7	Belgium	20,941
May 12	South Africa	22,788
May 16	New Zealand	35,761
May 26	Mexico	26,332
August 3	Republic of Ireland	37,040
August 29	Portugal	49,504
September 3	Portugal	19,600
October 3	South Korea	30,071
October 6	South Korea	33,027
November 7	Sweden	20,903
November 10	Costa Rica	12,914

5. Another way to relax is to go to the movies. Hollywood speaks about the highest grossing movies of all time, but as ticket prices go up, you don't have to sell that many tickets in 2021 to outperform a movie from the 1940s. What happens when you adjust for inflation? Well, here are the top 10 inflation-adjusted box-office blockbusters. Sum the number of tickets sold, and the inflation-adjusted dollars. The list goes from worst to first (data source: https://www.cnbc.com/2019/07/22/top-10-films-at-the-box-office-when-adjusted-for-inflation.html).

[20] Lawrence Dockery (2019, December 25) "USWNT average attendance 8% greater than US men's soccer team in 2019." https://worldsoccertalk.com/2019/12/25/uswnt-average-attendance-8-greater-us-mens-soccer-team-2019/ (accessed November 3, 2020).

Movie Name	Tickets sold	Dollars
	(Millions)	(Millions)
Snow White (1937)	109	982
The Exorcist (1973)	116.5	1,040
Doctor Zhivago (1965)	124.6	1,120
Jaws (1975)	128	1,150
The Ten Commandments (1956)	131	1,180
Titanic (1997)	143.5	1,290
E.T. the Extraterrestrial (1982)	147.9	1,330
The Sound of Music (1965)	157.2	1,410
Star Wars (1977)	178.1	1,600
Gone with the Wind (1939)	201	1,810

6. In the United States, baseball's nickname is "America's pastime." The ugly truth is that until October 1945, when Jackie Robinson signed to play for the Brooklyn Dodgers (he wouldn't play in a game until 1947), baseball was segregated. African American athletes played in the so-called Negro Leagues.[21] In one of the early years, 1921, the teams of the Negro National League won the following number of games (see Peterson, p. 257):

Chicago American Giants	41
Kansas City Monarchs	50
St. Louis Giants	33
Detroit Stars	30
Indianapolis ABCs	30
Columbus Buckeyes	24
Cincinnati Cuban Stars	23
Chicago Giants	10

In baseball, games are either won or lost, there are no draws or ties. So, by adding up the number of victories, how many games were played that season?

7. It's good to be king. Here are the world's top 10 palaces, by property values (as per https://www.lovemoney.com/gallerylist/56266/how-much-the-worlds-most-valuable-palaces-are-worth). What's their combined value?

---

[21] Their history is brilliantly told in Robert Peterson's classic book, *Only the Ball Was White* (New York: McGraw Hill, paperback edition 1984).

Palace	Property value (millions of pounds sterling)
Blenheim Palace (UK)	180
Windsor Castle (UK)	180
Ak Saray Palace (Turkey)	505
Istana Nurul Iman Palace (Brunei)	2,302
Buckingham Palace (UK)	3,840
Winter Palace (Russia)	4,900
Tokyo Imperial Palace (Japan)	9,300
Louvre Palace (France)	35,000
Palace of Versailles (France)	39,000
Forbidden City (China)	54,000

8. Legal battles can be expensive. Here are a few of the costliest settlements, many involving celebrity divorces. Estimate, then calculate, the total amount of the judgments (source: https://worthly.com/most-expensive/12-expensive-lawsuit-settlements-ever/).

Litigants	Settlement (millions of dollars)
Ralphs v. Six female employees	30
Ashley Alford v. Aarons	95
Bank of America v. George McReynolds	160
Michael and Juanita Jordan	168
Mercy General Hospital v. Ani Chopourian	200
Novartis v. Female Employees	250
Countrywide Financial Corporation	335
Tiger Woods v. Elin Nordegren	750
Rupert Murdoch v. Anna Torv	2,000
Dmitry Rybolovlev v. Elena Rybolovlev	4,500

9. The late Bill Shankley, former manager of Liverpool, once said "Some people believe football is a matter of life and death, I am very disappointed with that attitude. I can assure you it is much, much more important than that." In the United States, football is American football, and each team has a punter whose task is to get the team out of trouble by kicking the ball as far downfield as possible. Here (using the modern team names) is the list of top 12, all-time longest punts in the National Football League (NFL). What is the total yardage

covered by these punts? (Source: https://www.pro-football-reference.com/leaders/punt_long_single_season.htm)

Name	Team	Punt (yards)	Year
Steve O'Neal	New York Jets	98	1969
Shawn McCarthy	New England Patriots	93	1991
Randall Cunningham	Philadelphia Eagles	91	1989
Don Chandler	Green Bay Packers	90	1965
Rodney Williams	New York Giants	90	2001
Luke Prestridge	New England	89	1984
Bob Waterfield	LA Rams	88	1948
Dave Finzer	Chicago Bears	87	1984
Bob Scarpitto	New England Patriots	87	1968
Larry Barnes	San Francisco 49ers	86	1957
Bob Waterfield	LA Rams	86	1947
Sammy Baugh	Washington	85	1940

10. American football is extremely popular, at least in the US! So much so that the Super Bowl routinely ranks among the most-viewed television shows in the United States. One measure of popularity is a show's rating, which is the percentage of TV sets in the country that tune in to that particular episode. Here are the 10 TV shows in America that, so far, have had the highest ratings. What's the average?

M*A*S*H (the final episode, "Goodbye, Farewell, and Amen."	(1983)	60.2
Dallas ("Who done it?" As in, who shot J.R.?)	(1980)	53.3
Roots (Part VIII)	(1977)	51.1
Super Bowl XVI (The 49ers edged the Bengals, 26–21)	(1982)	49.1
Super Bowl XVII (Washington beat the Dolphins, 27–17)	(1983)	48.6
Olympics ice skating (Nancy Kerrigan and Tonya Harding)	(1994)	48.5
Super Bowl XX (The Bears drubbed the Patriots, 46–10)	(1986)	48.3
Gone with the Wind (Part One)	(1976)	47.7
Super Bowl XLIX (The Patriots pipped the Seahawks, 28–24)	(2015)	47.5
Gone with the Wind (Part Two)	(1976)	47.4

And here, without verbiage, are some sums to sum.

11. 98 + 42 + 69 + 27 + 93
12. 93 + 35 + 6 + 45 + 13
13. 16 + 68 + 84 + 51 + 19
14. 57 + 49 + 19 + 36 + 86

15. $13 + 3 + 32 + 61 + 82$
16. $9 + 30 + 8 + 18 + 98$
17. $59 + 27 + 28 + 57 + 68$
18. $41 + 77 + 48 + 13 + 21$
19. $84 + 29 + 81 + 72 + 90$
20. $50 + 45 + 57 + 12 + 37$
21. $91 + 99 + 96 + 69 + 8$
22. $46 + 45 + 21 + 51 + 64$
23. $50 + 12 + 97 + 10 + 19$
24. $1 + 51 + 28 + 78 + 42$
25. $97 + 26 + 79 + 40 + 3$
26. $30 + 76 + 64 + 90 + 73$
27. $68 + 62 + 83 + 99 + 96$
28. $40 + 97 + 73 + 99 + 41$
29. $84 + 72 + 86 + 53 + 9$
30. $49 + 49 + 70 + 89 + 26$
31. $10 + 17 + 80 + 15 + 59$
32. $45 + 85 + 75 + 83 + 32$
33. $92 + 46 + 71 + 71 + 64$
34. $78 + 58 + 83 + 93 + 20$
35. $8 + 18 + 64 + 99 + 85$
36. $88 + 42 + 43 + 98 + 84$
37. $68 + 20 + 55 + 37 + 93$
38. $23 + 96 + 17 + 7 + 90$
39. $35 + 35 + 15 + 2 + 90$
40. $33 + 32 + 57 + 13 + 71$
41. $26 + 88 + 67 + 64 + 20$
42. $23 + 38 + 86 + 43 + 71$
43. $99 + 39 + 92 + 17 + 71$
44. $11 + 88 + 89 + 29 + 90$
45. $56 + 98 + 19 + 96 + 19$
46. $35 + 34 + 70 + 28 + 31$
47. $58 + 65 + 7 + 95 + 43$
48. $65 + 94 + 72 + 49 + 97$
49. $20 + 94 + 22 + 48 + 86$

Interlude II

Checking, Check Digits, and Casting Out Nines

Die ganzen Zahlen hat der liebe Gott gemacht, alles andere ist Menschenwerk. (God made the integers, all else is the work of man.)

LEOPOLD KRONECKER

Master, we saw one casting out devils in thy name, and he followeth not us.

MARK 9:38 (KING JAMES VERSION)

On the back of this book is its International Standard Book Number (ISBN). This is a string of 13 digits—possibly with the capital letter X at the end instead of the 13th digit. The first three digits, the prefix, are 978, which say "this is a book." Then follows the "registration group element," a mishmash that specifies, among other things, the language of publication—0 or 1 for English, for example. The next five numbers, 19879, represent the registrant element, flagging that the publisher is Oxford University Press—the largest university press in the world. The next three digits are the publisher-specified numbers that identify this particular book, edition, and format. In countries that don't publish as many books as the United States or the United Kingdom, there can be a far longer registration group element. For example, all books published in Bhutan—the kingdom of the peaceful dragon—begin with the eight-digit sequence 97899936, with only the next four digits specified by the individual publisher.[1]

---

[1] Sonam Kinga, "Publications in Bhutan since the establishment of the ISBN agency," *Journal of Bhutan Studies*, 5 (2001), 78–93.

The last number is key: it's a check digit. That way, if something goes wrong in the scanning process (if you bought this in a bookstore), the machine would know—and if all is well, the bookseller's computer automated system knows to place an order with the publisher, a great benefit in this age of just-in-time inventory.

Let's use "mystical math" to find a check digit. The ISBN of *The Physics of Rugby*[2] has the first 12 digits 9781904761174. To calculate the check digit at the end, do a weighted sum. Multiply the digits alternately by 1 and 3. The weighted sum for the ISBN is $(9 + 8 + 9 + 4 + 6 + 1) + [3 \times (7 + 1 + 0 + 7 + 1 + 7)]$.

Rewrite this first term "mystically" as:

$$9 + 8 + 9 + 4 + 6 + 1 = 9 + 8 + (4 + 6) + [9 + 1] = 37$$

The second term contains:

$$7 + 1 + 0 + 7 + 1 + 7 = 23$$

The total comes to $37 + (3 \times 23) = 37 + 69 = 106$

The last digit in this case is 6. To find the check digit, subtract this number from 10. In this example, the check digit is $10 - 6 = 4$. The full ISBN is therefore 9781904761174. A laser scan of the barcode, if something goes wrong, will lead to a different check digit, and the book will need to be rescanned or inspected for problems. And now you can check the ISBN for *this* book, 9780198852650!

Anyone involved in selling things knows the importance of avoiding errors, hence the check digit. Mathematics, though, comes with its own check-digit system—casting out nines. It's a speedy way to check whether a sum, subtraction, or multiplication is even plausibly right—without doing the calculation in full. This can be of use in multiple-choice exams, as it may allow the rapid removal of some of the options proffered as answers. Luckily, finding the check digit for a number is far simpler than calculating the check digit of an ISBN. And casting out nines is of historical interest: it was included in the first-ever mathematics book to be printed in the West, the so-called *Treviso Arithmetic*, of 1478,

---

[2] A great book, but I'm biased. Trevor Davis Lipscombe, *The Physics of Rugby* (Nottingham, UK: Nottingham University Press, 2009).

a book that originally sold for 15 solidi, roughly the same price as three chickens.[3]

As an example, let us return once again to Al Capone and tax evasion. Earlier, we looked at the taxes he owed. Now look at his taxable, albeit ill-gotten, income (in US dollars).

Year	Income
1924	123,102
1925	257,339
1926	195,676
1927	218,056
1928	157,203
1929	103,999
	1,055,375

Is this total believable? The way to check is to cast out nines. Take the first year of income, $123,102. Add up the digits, which is 9. This is the digit sum. We care not one jot about the sum, only the remainder when dividing by 9, which is 0. Crossing out the digits that sum to 9, we can form the digit sum of his income for the year 1925 as 2̶5̶7̶, 339, which is 11, whose own digit sum is 2. For the year 1926, we see that 1, 5, 6, and 6 sum to 18 and so can be ignored, leaving a digit sum for his income of 7. Likewise, in the year 1927, ignore the 1 and 8 to get a digit sum for income of $2 + 5 + 6 = 13$, which has its own digit sum of 4. For the year 1928, note that 7 and 2 sum to 9, as do 1, 3, and 5, giving a check digit of income of 0. Last, in the year 1929, we can see straight away the digit sum for Capone's income is 4. The combined digit sums of the addends are $0 + 2 + 7 + 4 + 4$. The digit sum for this is 8. Now look at the proposed answer. The digit sum of 1,055,375, using the "mystical method" is $1 + (5 + 5) + (3 + 7) + 5 = 26$, which also has a digital sum of 8. As the digit sums match, we now know the answer is believable. Believable, but not necessarily correct (although, in fact, it is). Any mathematical anagram, a permutation, of the digits, such as 1,053,575 would generate

---

[3] The first arithmetic book printed in the Americas was the *Sumario Compendioso*, written by Juan Diez and printed in 1556 by Juan Pablos in Brescia, Mexico. It is a handbook of practical mathematics for calculating percentages, currency conversions, and dealing with the purity levels of gold and silver together with "some rules touching on arithmetic." See Bruce Stanley Burdick, *Mathematical Works Printed in the Americas 1554–1700* (Baltimore, MD: Johns Hopkins University Press, 2009).

the same check digit. Should the check digits not match, the answer *cannot* be correct. And, while this is an easy, elegant, and swift way to check a calculation (it works for multiplications as well), there are those who don't like it—including one author of a book published in the year 1716 (a year whose own digit sum is 6).[4]

[4] Edward Cocker, *Cocker's Arithmetic* (1716). For an extended discussion, see Maxim Bruckheimer, Ron Ofir, and Abraham Arcavi, "The case for and against casting out nines," *For the Learning of Mathematics*, 15(2) (1995), 23–8.

4

Quicker Quotients and Pleasing Products: Multiply and Divide by Specific Numbers

There is a divinity in odd numbers, either in nativity, chance, or death.

WILLIAM SHAKESPEARE,
Merry Wives of Windsor, Act V, Scene I

All Gaul is divided into three parts. (Gallia est omnis divisa in partes tres.)

Julius Caesar, *De Bello Gallico*

[Homer] is the father of arithmetic, because by saying that fifty men guarded each of the Trojan fires, he does not compute himself, but furnishes the occasion of computing the Trojan army at fifty thousand men.

Edward Gibbon, *The Miscellaneous Works of Edward Gibbon*
(John, Lord Sheffield, Ed.) London B. Blake, 1837.

Marin Mersenne, a French monk born in 1588, was skilled at mathematics. An entire class of prime numbers is named in his honor. These Mersenne primes are of the form $2^n - 1$. These aren't always primes (when $n = 4$, it generates 15, which most certainly isn't prime), but it's a great formula with which to hunt for large prime numbers, which are important in cryptography, keeping the nation and our bank account details safe. But is $2^{67} - 1$ prime? Mathematicians thought it might not be. On October 30, 1903, mathematician Frank Nelson Cole wrote down this number, 147,573,952,589,676,412,927, on the chalkboard. Then, with perhaps a flair for the dramatic, he wrote $= 193,707,721 \times 761,838,257,287$. The rest of his lecture—which ended with a standing ovation—consisted of multiplying the two numbers on the right-hand side to show that they indeed generated $2^{67} - 1$, thereby proving it wasn't a prime number. Cole, and possibly his audience, may have yearned for some techniques from rapid math to help hurry the proof along!

There are large numbers of tips and tricks to help multiply and divide numbers rapidly though, to be fair to Cole, they tend not to involve 9- or 12-digit numbers. The difficulty is organizing them, especially as some numbers have a similar trick to help things go quicker. This chapter is ordered by the numbers themselves, figuring that it's easier to look up the number you want than to try and remember a particular technique. Please, don't forget to look at a problem first to find a swift way to do it; by thinking first, you might invent a new method.

The famous English artist George Stubbs has an enormously famous—and famously enormous—painting, "Whistlejacket," hanging in the National Gallery, London. It measures some 115 inches × 97 inches. To calculate the area of the painting, there's no need to strive to find a technique to involve three-digit numbers ending in a 5, or numbers close to 100. The simple thing to do is to write the area as (115 × 100) − (115 × 3). The first you can write down straight away, 11,500 and the second is 345. Subtract (take off 350 to get 11,150 and then add the 5 back on) to get 11,155 square inches.

The secret is to look for, or invent, your own shortcuts, as well as learning the established tricks of the trade.

Multiply or divide by 4

The witches in Shakespeare's play *Macbeth* mutter "Double, double, toil and trouble." They were wrong. You can save yourself much toil and trouble if, instead of multiplying by 4, you simply double then double again.

To find 137 × 4, double the 137 to get 274. Double once more to get 137 × 4 = 548. This process, mentioned earlier, shows a hallmark of quick calculations: it is sometimes easier to do two simple steps (in this case, doubling twice) than doing one complicated step, multiplying by 4.

As another example, think of 23.4 × 0.4. As a rough estimate, this is about 20 × 0.5, which is 10. Ignore the decimal points, double 234 to get 468, then double once again to get 936. Inserting the decimal point to get an answer of about 10 means we have found 23.4 × 0.4 = 9.36.

Fans of soccer might find this double and double again strategy of use. In the English Football League Championship, for example, there are 24 teams. The guideline to get promotion is to win at home and to draw away. With 23 other teams in the league, you play them once at home—to earn a win worth three points—and once away, bringing home the single point from the draw. If you do so, you get 23 lots of 4 points, giving 46×2, which is 92. This is great news: the number of points you need for promotion is precisely twice the number of games you'll play. Those who enjoy non-league football might go to games in the National League South—featuring my former hometown team, Tonbridge Angels. With 22 teams in the conference, 88 points should give you a great chance at spending next season in the National League, which is exactly the number of points that Torquay United earned when they won promotion at the end of the 2018–19 season.

To divide by 4, all you have to do is to halve, then halve again.

So how do you convert inches to centimeters? With the new double-double and halve-halve strategy, the answer is "swiftly." One inch is 2.54 cm. If you want to know how many centimeters there are in 1 yard, recall there are 3 feet in a yard and 12 inches in 1 foot. The answer, then, is 36×2.54 cm. This might not look like multiplication or division by four, but it is both. The trick is to spot that $2.54 = 2.5 + 0.04 = 10/4 + 4/100$. To find our answer, then, we divide 36 by 2, twice, which is 9. We expect $(10/4) \times 36$ to be slightly more than 2×36, this gives 90. The second step is to double 36 to get 72, and double again, to get 144. As we expect $(4 \times 36)/100$ to be about 1, this part of the equation must have the value 1.44. Combined, we get 91.44 cm in 1 yard, which is the answer the calculator gives. Explaining this takes time; doing the calculation itself is quick.

In Islam's Abbasid Caliphate, the legal cubit was approximately 19 inches, the same value given for the biblical cubit by Rabbi Avraham Chaim Naeh (1890–1954). How many centimeters are there in Naeh's biblical and the Caliphate's legal cubit? Try it!

The cubit, the distance from the tip of the elbow to the tip of the middle finger is, like many early units, based on human dimensions. In 2012, the United States Army used beam calipers to measure what they called the forearm-hand length—the cubit—of 4,082 male military personnel, giving the (male) cubit as 48.02158 cm (about 18.9 inches), and 1,986 female military personnel, whose (female) cubit is 43.986 cm.[1] These combine to give a modern military cubit of 46.7 cm (18.3858 inches).

[1] The standard deviation for the men is 2.328644 cm, for the women 2.3376 cm. The data are available at Penn State (n.d.) "ANSUR II." https://www.openlab.psu.edu/ansur2/ (accessed August 3, 2020).

And there is another mathematical reason to like four. The word, written out, has 4 letters. It is the only number that equals the number of letters in its name!

Multiply or divide by 5

Multiplication

As with 4, it borders on insulting to mention multiplication by 5. But, if you think beyond something as simple as 12×5, such as 246×5, the multiplication becomes a good deal more complex. Or does it? Again, a standard trick in quick calculations is to multiply by 1 or, rather, to multiply by n then divide by n, for some suitably chosen n. For the number 5, life is sweet when $n = 2$. Namely, write:

$$246 \times 5 = 246 \times 5\frac{2}{2} = 246 \times \frac{10}{2}$$

The rule for multiplication by 5, then, is this: estimate it; halve it; append some zeros or insert a decimal point.

Estimate that 246×5 is more than $200 \times 5 = 1,000$. Halve 246 to get 123. To get an answer close to 1,000, glue a 0 on to the end to get $246 \times 5 = 1,230$.

To find 0.7×0.05, estimate the answer as less than $1 \times 0.05 = 0.05$. Halve 7 to get 3.5. Reinserting the decimal point to obtain an answer of the correct order of magnitude gives $0.7 \times 0.05 = 0.035$.

Division

This is not particularly surprising, given that you know the secret to multiplying by 5. If we puzzle over 243/5, do the same as before, multiply top and bottom by 2.

$$243 \div 5 = \frac{243}{5} \times \frac{2}{2} = 243 \times \frac{2}{10}$$

The insight: double the number, then add zeros or decimal points as necessary.

The result of 243/5 is slightly less than $250/5 = 50$. Take 243, double it to get 486, and now insert a decimal point to produce $243/5 = 48.6$.

Number buffs will recognize the strategy here of trying to find numbers that multiply or divide to give 10, or something equally easy to play with. This is a method used often in this book, and wherever quick calculations are discussed.

Multiply or divide by 6

There's no new tip here. Just remember to multiply by 3 and double your answer instead of multiplying by 6. You can order the operations differently, if you like, by doubling and then multiplying by 3. That said, recall that doubling is easier than almost anything else, so it's probably swifter to multiply a small number by 3 and then double rather than to double first, get a large number, and then try to multiply it by 3.

For those who love number factoids, 6 happens to be the smallest perfect number. It is divisible by 1, 2, and 3, which are its positive divisors, excluding the number itself. But $1 + 2 + 3 = 6$, and the definition of a perfect number is one that equals the sum of its positive divisors (the next one is $28 = 1 + 2 + 4 + 7 + 14$; after that, there's a big jump to 496 and 8,128).

Multiplication

As an example, consider 216 (which is 6^3). To find 6^4 we form 216×6. An estimate is $200 \times 6 = 1,200$. Take the 216 and triple it, which is 648, then double to get $216 \times 6 = 6^4 = 1,296$.

Another example: 19.3×0.6 is approximately $20 \times \frac{1}{2} = 10$. Triple 193 to get 579 and then double the result, which gives 1,158. Inserting a decimal point to give an answer of about 10 yields $19.3 \times 0.6 = 11.58$.

But wait—there's more! If you don't like the idea of tripling then doubling, use basic math to recall $6 = 5 + 1$. To multiply 216×6, multiply the 216 by 5—using the rapid method outlined in the previous section—to get 1,080, then add on another 216 to get 1,296. For 19.3×0.6, write down 9.65, add 1.93 to get 11.58.

Use the triple-double or five-plus-one method, whichever one gives you the answer most quickly. Pro tip: if you multiply any *even* number by 6, the last digit of the answer will be the same as the last digit of that other number. In other words, 128×6 must end in an 8. The multiplication $13,472 \times 6$ must end in a 2. It's a quick check on whether your answer is, at least, plausible.[2]

---

[2] If you write the even number as $10n + 2m$, then multiply by 6, you obtain $6(10n + 2m) = 10(6n + m) + 2m$. Hence, the last digits are the same, in this case, $2m$.

There is a practical advantage for multiplying by 6 quickly, especially if you live in, or are visiting, the United States. The most common sales tax rate among the 50 states is 6%. Multiplying by 6, then, will let you work out the tax you should be paying before it rings up at the cash register.

Using your newfound swiftness, multiply $12 \times 30.5 = 6 \times 61$ to reveal the answer of 366. The number 30.5 is the average of 30 and 31, which means that if we had 12 months, with the odd numbered months having 31 days, and the even numbered months having 30, we could have a year of 366 days. Strip one day from the last month of the year, which is our December, and you have a clean and simple calendrical system. Except, as the fascinating story of the calendar shows all too well, there is no clean and simple system: 365 and a bit solar days in a year, combined with 29 and a tad of solar days between new moons, produces a dazzling and baffling array of ways to track the passing of time.

Division

To divide by 6, take your pick. Divide by 2 and then by 3, or vice versa. If the number is even, it might be best to divide by 2 first, as this will result in a whole number to be divided by 3. If it's odd, you might divide by 3 first, assuming that if the result is nasty, it will be relatively easy to halve the result, as opposed to beginning with a fraction that then needs to be divided by 3.

Multiply or divide by 7

Multiplication

The number 7 is annoying. It's a favorite with teachers, usually because it's rather a pain to deal with. There's no wonderful way to ease the pain. But one thing you might try is this: remember $7 = 5 + 2$.

To find 234×7, write it as $234 \times 7 = 234 \times 5 + 234 \times 2$.

As a rough estimate, the answer is about $200 \times 7 = 1,400$. Now use the trick for multiplication by 5 (multiply by 10 and halve) and add it to double the number.

That is to say, halve the number and put a zero on the end. Then add twice the original number. Half of 234 is 117, which with a zero at the end becomes 1,170. Twice 234 is 468, and we add this to the 1,170 to produce $234 \times 7 = 1,170 + 468 = 1,638$.

This allows you to do a week-related problem. If you wanted to know the number of seconds in 6 weeks, you know that there are 7 days in a week. But there are 24 hours in a day, and $24 = 8 \times 3$. There are 60 minutes in an hour, which is $5 \times 6 \times 2$. And 60 seconds in a minute $= 10 \times 6$. That means there are $6 \times 7 \times 8 \times 3 \times 5 \times 6 \times 2 \times 10 \times 6$ seconds in a week. Shuffle, to get $10 \times 8 \times 7 \times 6 \times 6 \times 6 \times 5 \times 2 \times 3$. That gives $10 \times 8 \times 7 \times 6 \times 6 \times 6 \times 5 \times 3 \times 2$. As $6 \times 6 = 9 \times 4$, this means there are $10 \times 9 \times 8 \times 7 \times 6 \times 5 \times 4 \times 3 \times 2 \times 1$ seconds in 6 weeks, which is usually written 10!, where ! is pronounced "factorial" or, if you prefer, "bang!" Evaluating it, $10! = 3,628,800$, though you might want to wait until the end of the book to find a quick way to do this.[3]

Division

This is not a method, more of a neat party trick. The fun mathematical fact about 1/7 is that expressed as a decimal:

$$\frac{1}{7} = 0.\overline{142857}$$

As a reminder, the overbar means the digits over which the bar appears are repeated indefinitely. So 0.142857142857142857. . .

This may not seem particularly useful. The odd thing, though, is that if you go to the trouble of memorizing this sequence 142857, then you can immediately write down all the other fractions involving 7. The way to remember it is that 7, the number we seek to divide by, when doubled is 14, which when doubled is 28, which when doubled is 56, and when written together, that's 142856, which is one less than you need to remember. All you do is to start them off at a different point within the sequence.

$$\frac{2}{7} = 0.\overline{285714}$$

$$\frac{3}{7} = 0.\overline{428571}$$

$$\frac{4}{7} = 0.\overline{571428}$$

$$\frac{5}{7} = 0.\overline{714285}$$

$$\frac{6}{7} = 0.\overline{857142}$$

[3] If you want to make a habit of quick math, the monosyllabic "bang!" takes less time to say than the four-syllable "factorial."

With a rhetorical flourish of the pencil, chalk, or marker pen, you can impress friends and colleagues with your knowledge of recurring decimals. All you need to remember is 142857.

As an aside, recall that an often-used approximation for π is $22/7 = 3\frac{1}{7}$. We can now write this down immediately as 3.142857. However, $\pi = 3.1415926\ldots$ and so this approximation goes wrong fairly quickly.

As:

$$\frac{1}{7} = 0.\overline{142857}$$

This unique decimal expansion does allow a quick way to divide by 7. Suppose you seek 123/7. Estimate, by saying 123 is just under 140, so the answer must be less than 20. To begin, multiply 123×7 to obtain 861. Double, to get 1,722. Double again, to get 3,444 and shunt this two places to the right. That is to say, write:

$$
\begin{array}{cccc}
1 & 7 & 2 & 2 \\
  & 3 & 4 & 4 & 4
\end{array}
$$

Now sum to get 17.5644. However, we know the decimal expansions for fractions whose denominators are 7, the exact answer must therefore be $\frac{123}{7} = 17.\overline{571428} = 17\frac{4}{7}$.

How it works

We take the number, multiply by 7, and then double it. That means we are multiplying by 14. Doubling it again, we obtain 28. By shifting the numbers over, we are thus estimating multiplication by 0.1428, which is roughly the same as division by 7. However, as the decimal expansions are easy to memorize, we can replace our estimate with the exact answer.

As an added bonus, if you can remember the string of numbers 1, 3, 2, 6, 4, 5, you have a short-hand way to determine if a number is divisible exactly by 7. Suppose we want to see if 7,654,325 is divisible by 7. Write the number in reverse order, 5234567. Then split it up and multiply by the numbers in the string, repeating if necessary. In other words, form $(5 \times 1) + (2 \times 3) + (3 \times 2) + (4 \times 6) + (5 \times 4) + (6 \times 5) + (7 \times 1) = 5+6+6+24+20+30+7 = 98$. But 98 is divisible by 7, and so, therefore, is 7,654,325. This works because 1, when divided by 7, has a remainder 1. The remainder for 10 is 3; for 100 is 2; for 1,000 is 6; for 10,000 is 4; and for 100,000 is 5. The process is similar, then, to casting out nines. And to go even further, Leonardo of Pisa had a similar process of "casting out thirteens."

There is another test for divisibility by 7, which is perhaps more useful for smaller numbers. Consider 2,576. Remove the last digit, to form 257. Now subtract double the digit you've just taken off, in this case $2 \times 6 = 12$. Subtracting gives $257 - 12 = 245$. Now repeat. We form $24 - 10 = 14$. Once more gives $1 - 8 = -7$. As we have arrived at a single digit, albeit negative, we stop. As this single digit is divisible by 7, so is 2,576. The key here is that if you have an integer $10n + m$, say, then it has the same remainder when divided by 7 as the number $10n - 20m$. These two numbers differ by $21m$, which is an exact multiple of 7, and so the remainder of both numbers must be the same.

Multiply or divide by 8

As the Prime Minister often says during Britain's "Prime Minister's Question Time" in the Houses of Parliament, "I refer the Right Honorable member to the answer I gave some moments ago." When multiplying by 8, you double, then double, and then double again. To divide, you simply halve, halve again, and halve once more.

Another option is to recall basic arithmetic, given that $8 = 10 - 2$. If you wish to multiply 12.37×8, estimate as $12 \times 8 = 96$. Strip out the decimal points. Add a zero to the end of the number, to obtain 12,370. Double the number to get 2,474 and subtract. This you could do by writing it as $12,370 - 2,370 - 104 = 10,000 - 100 - 4 = 9,900 - 4 = 9,896$. Reinsert the decimal point to get a final answer 98.96.

While there's a pattern, it's always best to think. Books in the United States, for example, are not printed in metric sizes. One common book size, in inches, is $6\frac{1}{8} \times 9\frac{3}{4}$. If you need to know the surface area of the book, a rough estimate is going to be 6×10 or 60 square inches. To compute the actual value, one might choose to use improper fractions:

$$6\frac{1}{8} \times 9\frac{3}{4} = \frac{49}{8} \times \frac{39}{4}$$

This leaves a fun multiplication (49×39 drips with possibilities) and then you can use the rules for dividing by 8 and by 4. Lots of halving involved.

To calculate the answer more swiftly, though, recall what a fraction means. Namely, write:

$$6\frac{1}{8} \times 9\frac{3}{4} = \left(6 + \frac{1}{8}\right)\left(10 - \frac{1}{4}\right)$$

The second bracket is written as $10 - \frac{1}{4}$ rather than $9 + \frac{3}{4}$ because multiplying by 10 is easy, and easy calculations are what we thrive on. Expanding the brackets term by term gives:

$$6 \times 10 - \frac{6}{4} + \frac{10}{8} - \frac{1}{32}$$

So that:

$$6\frac{1}{8} \times 9\frac{3}{4} = 60 - \frac{12}{8} + \frac{10}{8} - \frac{1}{32} = 60 - \frac{9}{32} = 59\frac{23}{32}$$

As an aside, our beginning estimate of 60 square inches is within 0.5% of the actual answer, which is impressive. It's a glorious consequence of overestimating one number, underestimating the other, and having the errors in the estimates almost cancel each other out.

One reason you might want to know the surface area of a book is that you want to cover it or bind it. For centuries, leather was the material of choice to cover a book. Clearly, one would need twice the answer above (which is just the area of the front surface) but also would need to know the width of the book's spine. Assuming that's an inch, the book would need a minimum of 129 3/16 square inches of leather. There are even books that, gruesomely, are bound in human skin. There are 50 or so such volumes, and most have a connection to the medical profession. One of these is a famous anatomy book of 1543, Andreas Vesalius's *De humani corporis fabrica* (*On the Fabric of the Human Body*). The Royal College of Surgeons of Edinburgh has a notebook bound in skin from the notorious "resurrectionist" William Burke, who with William Hare, dug up bodies of the recently deceased and sold them to the local medical school for subsequent dissection.

Multiply by 9

To multiply by 9, don't. Recall that $9 = 10 - 1$. You need only do one calculation, shunt your answer over one place, and subtract. If you want to discover what 678×9 happens to be, write down $6,780 - 678 = 6,102$.

As a reminder, do this last subtraction swiftly as $6,000 + 780 - 678 = 6,000 + 102 = 6,102$.

The age-old mathematics classroom question is "when will I ever use this in the real world?" It's worth pausing, just before we reach double-digit numbers, to show that quicker-calculation skills can be put to good use. The United States has yet to turn fully metric, even though it signed the Treaty of the Meter (*Convention du Mètre*) in 1875. One consequence is the difficulty that Americans and Britons have in discussing the weather, for America clings to giving temperatures in Fahrenheit, while in Britain they are given in Celsius. To convert from degrees Celsius, C, to degrees Fahrenheit, F, use:

$$F = \frac{9}{5}C + 32$$

This is ideal for the techniques used so far. The hottest temperature recorded in Britain occurred in Faversham, Kent, which reached a blistering 38.5°C on August 10, 2003. To convert to Fahrenheit, multiply by 9. Or, rather, multiply by 10 to get 385, subtract 38.5 (or, better, subtract 35 to get 350, then take off a further 4 and add 0.5 to get 346.5). Multiplication by 9 is now done, and we need next to divide by 5. We know how to do this swiftly, and the answer that makes sense (as we need a realistic temperature in Fahrenheit) is 69.3. Last, add on the 32 (so, following two-step addition, add on 30 and then add on 2) to get 101.3°F.

The coldest temperature recorded in Great Britain was an icy −27.2°C, which occurred twice, both times in Scotland. First in Braemar, Aberdeen, on February 11, 1895, and second in Altnaharra, Sutherland, on January 10, 1982. To report this to an American, we'd form −272 and add 27.2. As you might expect, we'd add 22 to get −250, then add on the 5.2 to get −244.8. Dividing this by 5 (you know the trick!) gives −48.96 and now add 32 to get −16.96°F.

One issue with the conversion to Fahrenheit from Celsius is that the formula isn't particularly elegant. A reworking produces the prettier and more transparent expression:

$$9(C + 40) = 5(F + 40)$$

The clarity shows right away that the two temperature scales coincide at −40, that is to say, −40°C = −40°F.

Before moving on, one fun fact about the number 9 is that $(6 \times 9) + (6 + 9) = 69$. Not only that but, depending on the font you use or your handwriting, it's one of the few equations that looks the same upside down!

Multiply by 11

The number 11 has nothing to do with things divine, nor with the communion of heaven, nor is it a contact or a ladder to the higher world, nor has it merit at all.

Petrus Bungus, *Numerorum Mysteria* (*Mystery of Numbers*), "De Numero XI" (1591).

The number 11 might claim, along with the late comedian Rodney Dangerfield, that it "gets no respect." After all, a mathematics journal published an article entitled "11: The first uninteresting number?"[4] The trick used to multiply any two-digit number by 11 is, though, far from boring. In fact, it might be the quickest of the tips in this book.

Suppose you want to evaluate 63×11. Write down your number, in this case 63, leaving some space between its digits, 6__3. To find out the bit in the middle, simply add the two digits of the original number. Here it's $6 + 3 = 9$. In other words, $63 \times 11 = 693$.

It can get hairy when dealing with larger numbers. If you seek 97×11, then you have 9__7. In the middle, you have $9 + 7 = 16$. At this point, alarm bells ring, as the number is bigger than 10. Not to worry: carry the 1. That is to say, add the 1 to the 9 to get 10 and put the 6 in the middle. Thus $97 \times 11 = 1,067$. A fall back, for nastier numbers, is to bear in mind that $11 = 10 + 1$ and then do two simple multiplications and an easy addition. Therefore, 57.3×11 would be computed as $5,730 + 573 = 5,730 + 570 + 3 = 6,300 + 3 = 6,303$, and then put in the much-needed decimal point, to get $57.3 \times 11 = 63.03$ (I hope the removal of the decimal point—and the rapid estimate of the answer as $60 \times 10 = 100$ is, by now, second nature!)

The spacecraft Voyager 1, launched on September 5, 1977, hurtles through interstellar space—still transmitting useful scientific

---

[4] Eli Maor, "11: The first uninteresting number?" *Mathematics Teaching in the Middle School*, 7(5) (January 2002), 308–11.

information—whizzing through the outer reaches of the solar system, so NASA reports, at about 38,600 miles per hour. To put this into metric units, retain only the 386. We also need to know that 1 mph is roughly 0.00044 kilometers per second. It helps to put this into scientific notation. Voyager travels at 3.86×10^4 mph, and 1 mph is 4.4×10^{-4} km/s. We estimate that Voyager travels at about $3 \times 10^4 \times 4 \times 10^{-4}$, which is about 12 km/s. It remains to calculate 368×44, which we do by computing 386×11, and doubling the answer twice. That's $3,860 + 386 = 4,246$, which, when doubled gives 8,492, and when doubled again, is 16,984. As we expect an answer of about 12, this means the speed of Voyager I is 16.984 km/s.

Haribo sugar-free gummy bears have become an internet sensation. According to the nutritional information on the packet, a serving size consists of some 17 bears, which run to 41 g and contain 160 calories. Bags of 1 lb (453 g) can be purchased. Were anyone foolhardy enough to scoff the lot, how many calories would they consume? Well, each package contains 453/41 servings, which is about 11 servings. The total calorie content is going to be 160×11, which we know straight away to be 1,760 calories, which is not too far from the total amount of calories an adult should consume in a day. If reviews on social media are correct, though, you probably won't be too bothered about eating for a while after polishing off an entire packet.

How it works

Suppose the two-digit number is $10n + m$. When multiplied by 11, the product is:

$$(10n + m)(10 + 1) = 100n + 10n + 10m + m = 100n + 10(n + m) + m$$

In other words, the hundreds digit of the product is the same as the tens digit of the original number. The units digit is the same as the units digit of the original number. And the tens digit is the sum of the two digits in the original number.

Speaking of 11, there is another odd property worth noting. Take any two-digit number. Form another number by reversing the digits. Then add. The resulting number is always divisible by 11. For example, start with 58 and reverse the digits to get 85. In rapid-math style, their sum is $85 + 58 = 85 + 60 - 2 = 143$. And 143 is 11×13. If again we write the two-digit number as $10n + m$, reversing the digits gives $10m + n$. Adding yields $10n + m + 10m + n = 11n + 11m = 11(m + n)$, which means the answer is always a multiple of 11.

Multiply or divide by 12

When dealing with a dozen, it's sometimes best to split it up into $10 + 2$. Humans can double swiftly and multiply by 10 easily. You can compute 12×567 rapidly by adding a zero to get 5,670, and doubling the original number to get 1,134, and summing the rewards for your labors. The answer is $12 \times 567 = 5,670 + 1,134 = 6,804$. As before, you can do this summation rapidly by doing $5,670 + 1,134 = 5,600 + 70 + 1,130 + 4 = 5,600 + 1,200 + 4 = 6,804$. But this is a case when, if you have pencil and paper, it's easy to see how swift it can be. Immediately you write down, no need to think, 5,670. And as you write it down, you save some time by doubling the 567 in your head and write that down. Straight away, you have:

$$5,670+$$
$$1,134$$

This takes you about as long to write down as the traditional:

$$567\times$$
$$12$$

But the addition will be fantastically fast. To prove a point, think of a number with more digits, such as 68,937, and you'll see that the "$10 + 2$" method is quicker than doing the traditional multiplication.

To divide by 12, first divide by 3, then halve twice. In Ancient Egypt, instead of dividing the day into 24 hours of the same length, the period of daylight was defined to last 12 hours. That might be of practical value close to the Equator, but consider the case of Lerwick, the capital of the Shetland Islands, which lies about 475 km from the Scottish mainland, and about 540 km from Norway. June 21, 2020 saw a period from sunrise to sunset of 18 hours, 55 minutes, and 30 seconds, or 1,135.5 minutes. To convert this into Ancient Egyptian hours, we divide by 3 to get 378.5. Then halve to get 189.25, and halve again to obtain 94.625. Hence, in Lerwick in midsummer, an Ancient Egyptian hour would last 94 minutes, $37\frac{1}{2}$ seconds. On December 22, 2019, Lerwick saw only 5 hours, 49 minutes, and 15 seconds of daylight. That's $349\frac{1}{4}$ minutes. A swift way to do this is to spot that 348 is a multiple of three, and so to write it as $348 + 5/4$. Now divide by 3 to get $116 + 5/12$. Halve, which produces 58 5/24, and halve once again, giving a Lerwickian midwinter solstice Ancient Egyptian hour of 29 5/48 minutes, or a mere 29 minutes, $12\frac{1}{2}$ seconds. That's more than a modern hour shorter than the Ancient Egyptian hour for their midsummer solstice—daylight time flies by in the winter!

Multiply by 13

If you've tried to multiply by 13, you can understand better why triskaidekaphobia exists—it's a mess. Perhaps this is why bakers seldom give you a baker's dozen, 13, any more.[5] You can, though, use the same method that sometimes eases the pain with 7, and is the basis of multiplying swiftly by 12. Namely, recall that 13 = 10+3. So if you want do 45 × 13, say, you can add a zero to the number to make 450, triple the number, to get 135, and then add to get 45 × 13 = 450 + 135 = 585. Those who enjoy a little pizzazz in their calculations might want to try the quick method for multiplying by 52, and then halving the result twice. Good luck!

As all good mathematicians know, 13 = 11 + 2 = 12 + 1. The curious linguistic oddity about the number 13, though, is that the phrase "eleven plus two" is an exact anagram of "twelve plus one."

In spite of its arithmetical anagram, 13 is associated in Western culture with unfortunate things—many hotels don't have a 13th floor because of such superstitions. Nevertheless, 13 is one of the happy numbers. Happy numbers have an intriguing property. Take their digits one by one, square, and add them. Here, $1^2 + 3^2 = 1 + 9 = 10$. Now do the same again, which generates the number 1. Any number for which the sequence ends in the number 1 is a happy number. And if a number isn't happy, it will eventually settle down into a never-ending cycle, which would be bad luck indeed!

Multiply by 14

Just a friendly reminder: multiply by 7 and then double the answer. Or, if you prefer, write 14 = 10 + 4. Hence you can multiply by the number by 10, double the number twice, and add the two results. For

---

[5] Probably not. The baker's dozen dates back to medieval times, so it seems, when there were harsh laws for giving buyers short measure, possibly involving the removal of an arm or a leg from the baker. The thirteenth-century law *The Assize of Bread and Ales* set statutory sizes for some baked goods. Bunging a bun onto the original order guaranteed the baker could enjoy retirement with a full complement of limbs.

example, 58×0.14 is roughly $60 \times 0.1 = 6$. To do the "10 + 4" method, strap a zero to the end of 58 giving 580. Doubling 58 gives 116, and doubling 116 gives 232. Adding $580 + 232 = 812$ and so our answer, which must be approximately 6, is $58 \times 0.14 = 8.12$.

Another example lurks in the Rhind Papyrus, which is housed in the British Museum and which dates to 1650 BCE. Problem 69 requires readers to compute 80×14. We calculate 8×14, write it as $8 \times 7 \times 2 = 56 \times 2$ and we double this immediately and get our answer, $80 \times 14 = 1,120$. It's easier if you don't have to use hieroglyphics, or hieratic Egyptian script.

Shakespeare wrote sonnets, 154 of them in fact. A sonnet, such as his 18th ("Shall I compare thee to a summer's day") has 14 lines. To find out how many lines of poetry he wrote of this type, we need to compute 154×14. Add a zero to 154 to get 1,540. Double the 154 to get 308. Double again to get 616. The Bard of Avon therefore composed $154 \times 14 = 1,540 + 616 = 2,156$ lines in sonnet form.

Multiply or divide by 15

Multiplication

At the risk of repeating an idea too often, recall that $15 = 10+5$. Simply put, add a zero to the number, halve the result, and add them both. If you'd like to calculate 3.84×0.15, estimate the answer as about 0.3 and then consider 384×15. Add a zero to get 3,840, halve to get 1,920, and add to get 5,760. As the answer is about 0.3, we find that $3.84 \times 0.15 = 0.576$.

Life is full of choices, so you have another option. Because $15 = 30/2$, you could multiply the number by 3 and halve the result, adding zeros as necessary. Using the same example, we multiply 384 by 3 to get 1,152 then halve this, which is 576 and, putting the decimals in, $3.84 \times 0.15 = 0.576$.

Division

One glorious thing about the number 15 is that it is 10×1.5. And, speaking in fractions, $1.5 = 3/2$. Thus, dividing a number by 15 is the same as dividing the number by 3, doubling the result, and dividing that by 10.

For example, to calculate 348 ÷ 15, estimate it as more than 300 ÷ 15 = 20. Divide 348 by 3 to get 116. Double the result to get 232. To get an answer of about 20, we must have 348 ÷ 15 = 23.2.

The Arecibo radio telescope in Puerto Rico, which collapsed in 2020, stood 150 m in height. The Burj Khalifa skyscraper in Dubai is 829.8 m tall. Those who wonder how many Arecibo radio telescopes have to be placed one on top of the other to reach the height of the Burj Khalifa— surely you have, haven't you?—can now find the answer rapidly. To calculate 829.8/150, estimate as roughly 900/150 = 6, and note this is an overestimate. Divide 8,298 by 3, to get 2,766 and double to obtain 5,532, and so our answer is 5.532.

Multiply or divide by 15, 25, 35, or numbers ending in 5

Multiplication

As the numbers grow, different new techniques to handle multiplications and divisions become scarcer. But with numbers ending in 5, there is a fruitful method that is, like many others, fairly obvious—once it's been pointed out. We've seen it already, in multiplication by 15. Namely, as multiplying by 10 is easy, and so is halving, we write 15 = 30/2 and so multiply by 3 and halve. For 35, rewrite as 70/2, to multiply by 7 and halve.

One gift, though, is to multiply by 25. Suppose you want to multiply 264 × 25. To do so, multiply it by 1, which in this case is 4/4. Or, to be more explicit, we form:

$$264 \times 25 = 264 \times 25 \times \frac{4}{4} = (264 \div 4) \times 100$$

To multiply a number by 25, then, halve the number twice and add zeros or put a decimal point in the appropriate place. To estimate, we know that the answer has to be more than 200 × 25 = 5,000. We halve 264 to get 132 and halve 132 to get 66. Adding extra zeros to get the answer that agrees with our estimate leads to 264 × 25 = 6,600.

To compute 184 × 35, double the 35 to get 70. Perform the multiplication 184 × 70 = 12,880, and halve to get the result 184 × 35 = 6,440. Or, perhaps quicker, you could rewrite 184 × 35 as 92 × 70 = 6,440.

King Charles the Martyr lived somewhat lavishly. A report from 1681[6] claims that the royal household managed to consume each year "1500 oxen, 7000 sheep . . . 6,800 lambs . . . 470 dozen of hens . . . 1,470 of chickens . . . 364,000 bushels of wheat." Clearly, there were few, if any, vegetarians in the royal court. Focus on the wheat, though, which was used to make bread. The bushel is a measure of volume that equals 35.239 l, or approximately 35.25 l. To compute the amount of wheat in liters, we estimate it as roughly $30 \times 300,000$, or about 10,000,000 l. For higher precision, use the method for multiplying by 35 and then for multiplying by 25. Compute 364×35 in two steps. First, multiply the 364 by 7 to get 2,548, then halve to obtain 1,274. To get an answer of the correct size, we have to add a zero, so that $364 \times 35 = 12,740$. The next stage is to multiply 364×0.25, which we do by halving, to get 182, and halving again, to get 91. Summing the two gives $364 \times 35.25 = 12,740 + 91 = 12,831$. As we know the amount of grain has to be about 10 million liters, we have determined that each year, the court of King Charles I consumed bread made from 12,831,000 l of wheat.

The identical process holds for 45: multiply by 9, divide by 2, and insert decimal places or add extra zeros as necessary. This may have everyday use, or not, since one (Imperial) gallon is approximately 4.54609 l, which we can reduce to 4.545. Internet site Bklyner.com reported (November 20, 2014[7]) that the Verrazano-Narrows Bridge—the iconic structure linking Staten Island with Brooklyn[8] on which the New York City marathon begins—stood in need of repainting, and that 11,530 gallons of paint should do the job. In liters, this paint job would be roughly $12,000 \times 4.545$ l, which we can guesstimate as $12,000 \times 9/2 = 56,000$ l of paint. Armed with this fact, we form 115,300 and subtract 11,530, to get $115,300 - 11,530 = 115,300 - 11,300 - 230 = 104,000 - 230 = 103,770$. With multiplication by 9 completed, divide by 2 to get 51,885 l. This, though, gives us only a multiplication by 4.5. To add the extra 0.045, shunt the 51,885 over a couple of places to get 518.85 and add. Hence, it takes about $51,885 + 518.85 = 51,885 + 515 + 3.85 = 52,403.85$ l

---

[6] Tomas De-Laune, *The Present State of London*, printed by George Larkin for Enoch Prosser and John How, at the Rose and Crown. London, 1681, pp. 120–1.

[7] Mary Bakija (2014, November 20) "25 things you may not have known about the Verrazano-Narrows Bridge." https://bklyner.com/25-facts-verrazano-narrows-bridge-bensonhurst/ (accessed November 4, 2020).

[8] Technically it's between Richmond county and King's County.

of paint to cover the Verrazano-Narrows Bridge in its distinctive shade of gray.[9]
Consider the death of George, 1st Duke of Clarence, who died in the Tower of London in 1478. Tradition has it that he died in a butt of Malmsey, a story popularized in Shakespeare's play *Richard III*. A butt contains about 105 Imperial gallons of liquid—in this case Malmsey, a type of wine. Into how many of liters of wine was the good duke immersed? The conversion factor remains 1 Imperial gallon to 4.546 l, so we expect an answer of more than 400 l. Multiply 105 by 9 to get 945. Halve to obtain 472.5. That's our multiplication by 4.5 done, and so now we shunt our numbers over by two places to get 4.725 and add them, giving one butt of wine as about 477.2 l.

Naval hero Horatio, Lord Nelson died at the Battle of Trafalgar in 1805. In order to transport him home to Britain and keep his body in as good a condition as possible, the surgeon on board *HMS Victory*, Sir William Beatty, put Nelson's body in a barrel of brandy. It's not known whether he was inspired by the story of the Duke of Clarence or not. What is known is that, owing to chemical reactions that released gas into the brandy, the lid of the barrel burst, causing some of the sailors to think the beloved admiral had come back to life.

Division

Not surprisingly, to divide, do the opposite. Division by 25 requires an estimate, followed by doubling the number twice, and adding zeros. In one of the movie versions of King Kong, the giant gorilla was scaled to be 25 feet. In the 2005 movie, he climbs up the Empire State Building, which from floor to tip is 1,454 feet. To find how tall the Empire State Building is in terms of King Kong body lengths, we form 1, 454/25. This is about 1, 500/25 = 60. We double 1,454 to get 2,908 then double again, which is 5,816. Given our estimate, the Empire State Building is 58.16 King Kongs in height.[10]

To divide by 35, divide by 7 and double the result. To divide by 45, divide the number by divide by 3, divide by 3 again, and halve. Things get

---

[9] High marks go to those who know that $1/11 = .909$, so that $50/11 = 45.45$. A quick estimate is that the amount of liters you need is $11, 530/11 \times 50$. From there, divide by 11, then use the method for quick multiplication by 5.

[10] In *Skull: Kong Island* (2017), the gorilla was 104 feet tall, reducing the Empire State Building to a trifling 14 Kong lengths.

messy, rapidly! But with your estimate in hand, you know how many zeros to add or where to place the decimal points.

Multiply by 16, 26, 36, or numbers ending in 6

To multiply by 16, you can double four times in a row. And what better way to develop this skill than by mulling over the number of discarded wet wipes. The BBC reported on July 31, 2019 that over a 3-day period, some 16 metric tons of wet wipes accumulated in the drains of the city of Bristol. Suppose the local authorities wanted to find out how much that amounts to in a year. Well, as a leap year is 366 days, we would expect $16 \times 366/3$ metric tons to accumulate in that period, which is 16×122. Doubling once gives 244, twice is 488, three times is 976, and so our answer, obtained by doubling once more, is a startling 1,952 metric tons of wet wipes per year. To divide by 16, you can halve four times in a row.

For other numbers ending in 6, things are a bit trickier. The easiest thing to do might be to replace 76, say, by $75 + 1$, then use the quick method to multiply by 75 and then add in the number itself.

Multiply by 18, 27, 36, or multiples of 9

To multiply by 16, it's tempting to multiply by 9 and then double, or if you prefer, double, then multiply by 9. But there's a quicker alternative, based on shunting. Double the number. Shunt it one space to the right, then subtract. In other words, we are writing 18 as $(20 - 2)$. Consider 18×4.3. This is about $20 \times 4 = 80$. Double the 43 to get 86. By shunting one space to the right, it means we convert 86 into 8.6. Then do $86 - 8.6 = 86 - 9 + 0.4 = 77.4$.

But there's a bonus feature. Multiples of 9 all lend themselves to shunting. $27 = 30 - 3$, so multiply by 3 and shunt; $36 = 40 - 4$, so multiply by 4 and shunt. Have fun!

One piece of data that chemists find useful is the molecular weight of a particular compound. Consider, for example, Tobramycin, a drug specifically tailored to help those who suffer from cystic fibrosis. The formula for Tobramycin is $C_{18}H_{37}N_5O_9$. Every molecule of Tobramycin, then, consists of 18 carbon atoms (of atomic weight 12); 37 hydrogen atoms (of weight 1); 5 nitrogen atoms (of atomic weight 14); and 9 oxygen atoms (of atomic weight 16). The molecular weight of Tobramycin, then, is going to be:

$$(12 \times 18) + (1 \times 37) + (14 \times 5) + (16 \times 9)$$

Multiplying by 1 and by 5 is easy, so we know this sum is going to be:

$$(12 \times 18) + 37 + 70 + (16 \times 9)$$

To find (9×16), do the usual (multiply by 10 to obtain 160 and take 16 off, to give 144). To compute 12×18, we can use the method in this section, which is to double 12 to get 240 and take 24 off to obtain 216. Therefore, the molecular weight of Tobramycin is $216 + 37 + 70 + 144$. Using the "mystical" process, reorder this to get $200 + 100 + (30 + 70) + [16 + 44] + 7$. The total is 467.

There are other ways to find 12×18 quickly; use the method that is fastest and safest. This could be important if you plan to celebrate your newly developed rapid-math skills by ordering 12 Melchiors of champagne. A Melchior—named after one of the three Magi who, according to tradition, brought gifts to the baby Jesus—contains 18 l of bubbly. The total volume you have available is 12×18 l. An easy way to do this is to revert to your early days of learning your times tables. Namely, as $12 \times 9 = 108$, double this to $12 \times 18 = 216$. You could spot that $12 \times 18 = (15 - 3) \times (15 + 3)$ and use the difference of two squares to write this as $225 - 9 = 225 - 10 + 1 = 216$.

If you enjoy calculating molecular weights quickly, why not compute them for odd-sounding molecules like penguinone ($C_{10}H_{14}O$), whose molecular structure diagram looks like a penguin. Or else there's cadaverine, $NH_2(CH_2)_5NH_2$, a chemical produced by rotting tissue. Who says organic chemistry isn't fun?

Those who like a true challenge can try titin, a protein that happens to have the longest chemical formula, one that takes over 3.5 hours to read. Luckily, the short-hand form for the molecule is $C_{169723}H_{270464}N_{45688}O_{52243}S_{912}$. As the atomic weight of sulfur, S, is 32 you can now compute the molecular weight of titin, though I'd advise using a calculator in this case!

Multiply by 19, 29, 39, or numbers ending in 9

The technique here relies on the fact that it is always easy to multiply by a factor of 10. To multiply 58×39, say, you multiply $58 \times 40 = 2,320$ and then take 58 off. Or, rather, take off another 60 then add on 2. This gives $58 \times 39 = 2,320 - 60 + 2 = 2,262$. Along the same lines, should you prefer, multiply out the two brackets $(60 - 2) \times (40 - 1) = 2,400 - 60 - 80 + 2 = 2,400 - 140 + 2 = 2,262$. Extending the idea further, this will also work for 38, 58, 78... as you can use this method to calculate by 19, 29, 39... and double the result.

A villanelle is a particular form of poem that consists of 19 lines. A famous example is Dylan Thomas's 1951 poem "Do not go gentle into that good night." The book *Villanelles*[11] contains, at a quick count, 181 such poems. To work out the number of lines of poetry, we seek 181×19. Multiply by 20, which gives 3,620 and then take 181 off, which, naturally, means we subtract 180 to get 3,440 and then one more to get 3,439 lines of poetry.

In 2016, astronomers reported the discovery of five giant exoplanets.[12] One of them, labeled WASP-119b, reportedly has a mass equivalent to 1.2 Jupiters. The mass of Jupiter is approximately 1.9×10^{27} kg. Hence the newly discovered planet is $1.2 \times 1.9 \times 10^{27} = 12 \times 19 \times 10^{25}$ kg. And we know that $12 \times 19 = (12 \times 20) - 12 = 228$, and so the mass of WASP-119b is 2.28×10^{27} kg.

In the never-ending quest to make chemistry more appealing to students, in 2003 Chanteau and Tour created some ingenious organic molecules that look like stick figures. They are called NanoPutians, a pun on the Lilliputians, a race of small people, made famous in Jonathan Swift's novel *Gulliver's Travels*, published in 1726. The formula for one particular NanoPutian is $C_{39}H_{42}O_2$. We return, then, to the calculation of atomic weights, and rejoice that, as a NanoPutian has 39 carbon atoms, we can now use the method to calculate multiples of 39 swiftly. As the atomic mass of carbon is 12, of hydrogen 1, and oxygen 16, we know the carbon has mass $12 \times 39 = (12 \times 40) - 12 = 468$. We add 43 for the hydrogen, and then $2 \times 16 = 32$ for the oxygen. Hence, the molecular weight of this particular NanoPutian is 542.[13]

[11] Anni Finch and Marie-Elizabeth Mali (eds.) *Villanelles* (New York: Everyman's Pocket Library Series, 2012).

[12] NASA (n.d.) "WASP-199 b." https://exoplanets.nasa.gov/exoplanet-catalog/4671/wasp-119-b/ (accessed November 19, 2020).

[13] Stephanie H. Chanteau and James M. Tour, "Synthesis of Anthropomorphic Molecules: The NanoPutians," *Journal of Organic Chemistry*, 68(23) (2003), 8750–66.

Try these

1. Please don't let the extra zeros at the end, or the decimal points in front distract you. What's more, if you see something like 4.7 × 1.1 and puzzle over how to computer it, you can switch it around to get 1.1 × 4.7, which might be more easily recognized as a multiplication by 11. Bon appétit!
2. 4.20 × 0.38
3. 47 × 1.7
4. 1.8 × 4.2
5. 320 × 1.1
6. 0.23 × 2.50
7. 0.6 × 1.3
8. 1.5 × 4.7
9. 6.9 × 1.8
10. 2.8 × 82
11. 4, 300 × 0.21
12. 790 × 180
13. 1.6 × 250
14. 1.4 × 0.4
15. 1.1 × 5.8
16. 0.75 × 1.20
17. 0.11 × 1, 600
18. 1.2 × 90
19. 1.8 × 2, 100
20. 7.7 × 1, 300
21. 240 × 0.84
22. 180 × 0.016
23. 0.15 × 6.60
24. 5.2 × 1.7
25. 0.92 × 7, 100
26. 3, 000 × 0.5
27. 0.72 × 18
28. 1.2 × 3.7
29. 8.3 × 12
30. 0.12 × 280
31. 0.87 × 110
32. 1.30 × 0.14
33. 80 × 1.6

34. 160×4.3
35. 2.8×1.9
36. 4.8×4.1
37. 710×0.14
38. 1.50×0.77
39. 6.1×150
40. 1.9×1.7
41. $9.7 \times 5,000$
42. 110×6.1
43. 3.2×1.1
44. 210×5.8
45. 650×6.1
46. 0.014×3.400
47. 680×3.2
48. $1.8 \times 3,400$
49. 0.510×0.086
50. $1,100 \times 1.8$
51. 0.23×1.6

Multiply by 21, 31, 41, or numbers ending in 1

The number 41 seems plain. But when you write it out, something odd happens. In the phrase "forty-one," there is an odd combination. "Forty" is the only number whose letters are written in alphabetical order and, in contrast, the letters in the number "one" are in the exact opposite of alphabetical order. If that doesn't seem topsy-turvy enough, consider $7 \times 3 = 21$. Seventy-three, it so happens, is the 21st prime number. If you swap the numbers around, 37 is—you guessed it—the 12th prime. For yet more coincidences, 37, 73, 137, and 173 are all prime numbers, and as $137 \times 73 = 10,001$, you can divide by 137, approximately, by multiplying by 73 instead, provided you insert the decimal point in the right place. That can be useful in particle physics, as the so-called fine-structure constant α is $1/137$. These facts, though, don't help us multiply by numbers ending in 1.

To do so, you follow down the same primrose path as the previous section, but it is easier, as you add at the end rather than subtract. To calculate 58×21, multiply 58×20 (i.e. take 58, double it, and add a zero at the end) to get 1,160, then add another 58 to get $58 \times 21 = 1,218$

(again, you might prefer to add 60 and subtract 2, to get $1, 160 + 60 - 2 = 1, 220 - 2 = 1, 218$).

Writing in 50 BCE, Roman historian Sallust observed, "The poor Britons—there is some good in them after all—they produce an oyster." Whitstable, Kent, has long been a center for oyster fishermen, and the *Dictionary of Kentish Dialect*—compiled in 1888 by the Reverends W.D. Parish and W.F. Shaw—said that in Whitstable, oysters were sold by the prickle. A prickle, it turns out, is 10.5 gallons. If you purchased 6 prickles of Whitstable oysters, this would amount to $6 \times 10\frac{1}{2}$ gallons. To calculate this rapidly, double one and halve the other, to get $3 \times 21 = 63$ gallons. As these are "customary units," used by people in a particular trade, the oyster fishers would not have thought in terms of gallons. Two prickles is 21 gallons, which is a tub—also known as a London bushel. But, to make things more complicated, oysters were also sold in a Winchester bushel, which is 21 gallons, 1 quart, and $\frac{1}{2}$ pint.[14] Luckily, the metric system makes calculations far easier!

In 1547, King Edward VI commissioned an inventory of the property owned by his father, the late King Henry VIII. This included 120 firkins of saltpeter, known to modern chemistry as potassium nitrate. A firkin is approximately 41 l. King Edward VI therefore inherited some $(40 + 1) \times 120 = 4, 800 + 120 = 4, 920$ l of saltpeter, a basic ingredient of gunpowder. This might have been wise, given the precarious political position of his kingdom at that time.

Those seeking a challenge, ponder that most British of all sandwich spreads, Marmite. Each 100 g of yeasty goodness contains 39 g of protein, 29 g of carbohydrates, 3.1 g of fiber, and 6.1 g of salt. What's more, Marmite now comes in tubs of 600 g. You can use this section to find the protein, carbohydrate, fiber, and salt content of a 600-g tub quicker than you can say "Marmite lover."

Multiply by 32, 42, 52, or numbers ending in a 2

The method here is to decompose 72, say, into $70 + 2$. If you want 416×72, don't worry. Simply multiply by 7 and add a zero on to get 29,120. Then double 416 to get 832 and add, so that $416 \times 72 = 29, 952$. And, like worn-out comedians, we can recycle old material: if

---

[14] See "Prices and measurements of oysters in Whitstable," in J.E. Stevens, *Whitstable Native: A Short Study of Whitstable and Oysters* (Dartford, UK: Hartley Reproductions, 1977), p. 27.

the number in front of the 2 is even, you can use the techniques for numbers ending in a 1 and then double the answer.

Fifty-two is a gloriously composite number. While it may seem hard to multiply by 52, if you think of it as 50 + 2, your worries and fears subside. All you need to do is to multiply by 5 and 2, arguably the two easiest tricks in the book. For example, there's a popular card game known as Pounce, or Nertz. It's a fast-action game where cards end up all over the table as players jostle for position to lay down their last card and shout "Pounce!" If you have an enormous game with 17 people, what's the largest number of cards that can end up on the table? The answer is 17 × 52, but what's that? First, halve the 17 to get 8.5 and add 0 to get 85. That's mere multiplication by 5, so to multiply by 50, strap on another zero to get 850. Now double the 17 to get 34 and add to the 850 to get 884 cards. If you feel keen, you can add an extra 34 to account for the jokers.

Astute readers will see here an indirect, but possibly fast way to multiply by 13 and 26. That is to say, multiply by 52, as it's relatively swift, and then halve your answer to multiply by 26, or halve again if you seek to multiply by 13.

The practical value of computing multiples of 52 quickly is that it allows you to work out the number of weeks in a given number of years (then, multiply by 7, to find the number of days in that timeframe). In the calendar of the ancient Mayans, there was the Alautun, a period of approximately 63,081,429 years. To find the number of weeks in one Alautun, we divide this by two and multiply by 100, which gives 3,154,071,450. Doubling the original number yields 126,162,858, and add those two to get 3,280,234,308 weeks. If you want to find the days, you can multiply by 7, which as a reminder is 5 + 2, so you can use the same double and halve strategy as for 52. The answer is 22,961,640,156 days.[15]

Multiply or divide by 75

Multiplication

The trick here should be familiar. Namely, $75 = \frac{150}{2}$. Thus, use the trick to multiply by 15, add a zero on the end, and halve the result.

---

[15] The actual Alautun period is 2.304×10^{10} days, as the number of solar years is approximate.

Each year, the United States celebrates its independence on July 4. What better way could there be to mark America's freedom than by gorging on hot dogs? This is the day when Nathan's Famous Hot Dogs hosts an eating contest in Coney Island, New York. The habitual winner is Joey "Jaws" Chestnut, who as I write, is ranked number one by Major League Eating. His record, set in 2018, was to eat 74 hot dogs and buns (HDBs) in the allotted 10-minute time span of the competition. An HDB contains 297 calories, so how many calories did Mr. Chestnut consume in the contest? To compute this swiftly, set $297 \times 74 = (300 - 3) \times (75 - 1) = 300 \times 75 - (3 \times 75) - 300 + 3$. First, we compute $3 \times 75 = 225$. Strap on two zeros to get $300 \times 75 = 22,500$. Hence $297 \times 74 = 22,500 - 225 - 300 - 3$. From here, do whatever feels best. I'd write $225 + 300 = 500 + 25$ and then calculate $22,500 - 500 - 22 = 21,978$. This is a terrifying amount of calories, about 10 times the number you need to consume during a single day. Mind you, Americans like their HDBs—they eat an astonishing 18,000,000,000 of them every year.

Division

This is the same as dividing by 150 and doubling the result. And dividing by 150 is the same as dividing by 15 and then dividing by 10. So, the key is to remember the trick for dividing by 15, which is to divide the number by 3, and double the result. Confused? Don't be. To divide a number by 75, divide it by 3. Double the result. Double it again, and divide by 100.

Thus, to find $12,965 \div 75$, estimate it as $10,000 \div 100 = 100$, so we're looking for an answer in the hundreds. Divide 12,965 by 3, which is 4,321.66666. Double it, to get 8,643.3333. Double again to get 17,286.66666. Adding back in the decimal points, the answer that's in the hundreds is $12,965 \div 75 = 172.8666$.

Try these

1. 56×80
2. 440×240
3. 5.3×5.4
4. 2.1×9.6
5. 35×4.4
6. 820×0.32
7. 3.2×63
8. 8.3×9.7

  9. 49×97

10. 4.40×500

11. 680×7.7

12. 7.3×110

13. 0.84×7.20

14. 45×64

15. 730×7.7

16. 7.8×0.99

17. 3.9×9.3

18. 2.6×6.9

19. 63×3.2

20. 59×120

21. 270×250

22. 5.2×0.32

23. 9.1×75

24. 9.2×3.6

25. 4.7×190

26. 220×4.6

27. 8.5×8.1

28. 5.9×790

29. 680×0.95

30. 0.55×8.80

31. 780×2.7

32. 7.7×760

33. 0.83×3.5

34. 4.1×0.72

35. 0.52×0.83

36. 6.90×1.80

37. 370×49

38. 0.62×8300

39. 8.0×5.4

40. 770×0.68

41. 6.6×470

42. 850×6.2

43. 9.40×0.57

44. 8.8×8.1

45. 380×25

46. 5.2×790

47. 3.0×0.15

48. 76 × 62
49. 2.40 × 0.75
50. 0.056 × 1.300

Multiply by 111

Eleventy-one, a rather curious number, is fun to multiply. There are two categories. The first involves two-digit numbers. Let's look at 111 × 72. All you have to do is to add the two digits of 72, which gives 9. Insert this, twice, between the digits of the original number. That is to say, 111 × 72 = 7,992. There is a slight wrinkle, though, if you add the digits and they are greater than 10. In the case of 111 × 84, for example, 8 + 4 = 12. This means we have to carry the 1. More visually, we write the answer as:

$$
\begin{array}{cccc}
8 & 0 & 0 & 4 \\
1 & 2 & 0 & 0 \\
  & 1 & 2 & 0 \\
\end{array}
$$

And sum to get 111 × 84 = 9,324.

The second case involves triple-digit numbers, or higher: 111 × 723, for example.

The units column of the answer matches the units column of the multiplicand, thus 3. The tens digit equals the sum of the rightmost two digits, thus 2 + 3 = 5. The hundreds digit is the sum of all three digits 7 + 2 + 3 = 12 (so a 2, and we have 1 to carry). The thousands digit is the sum of the left-most two digits 7 + 2 = 9, which, with the one we carried, becomes 10, so thus a 0 and we have to carry a 1. The thousands column is the left-most digit, 7, plus the 1 carried, which makes 8. Thus 111 × 723 = 80,253.

If you prefer to do no work at all, apart from an addition, then you can simply shunt the numbers over. That is to say, you write:

$$
\begin{array}{ccccc}
7 & 2 & 3 &   &   \\
  & 7 & 2 & 3 &   \\
  &   & 7 & 2 & 3 \\
\end{array}
$$

Add the columns to get 80,253. You save time by not even bothering to add the extra zeros that you ought to have put in to make this a "real" sum.

One fun mathematical nugget is that $37 \times 3 = 111$. By shunting, you can write down straight away that $3 \times 37,037,037,037 = 111,111,111,111$.

There is another curiosity (well, not really). Because $111/3 = 37$, we can multiply the top and bottom of the fraction by 2 to get $222/6 = 37$. Or by 3, to get $333/9 = 37$. This is obvious, but it pleases the eye to write these as:

$$111/(1 + 1 + 1) = 37$$
$$222/(2 + 2 + 2) = 37$$
$$333/(3 + 3 + 3) = 37$$
$$444/(4 + 4 + 4) = 37$$

Or, more succinctly:

$$\frac{111}{1 + 1 + 1} = \frac{222}{2 + 2 + 2} = \frac{333}{3 + 3 + 3} = \frac{444}{4 + 4 + 4} = \cdots$$

Which is rather pretty. All we're doing, of course, is constructing multiples of 111, which are of the general form $100n + 10n + n$ and dividing them by $3n$. The n's cancel, leaving us with $111/3 = 37$).

Cricket fans may call the score of 111 a Nelson, after Admiral Horatio Nelson, famous for his three naval victories at the Battle of the Nile, Copenhagen, and Trafalgar. A Nelson, double Nelson, and higher multiples are supposedly unlucky. By the time you manage to score a quadruple Nelson, aka, a Salamander, you've chalked up 444 runs, which must surely look like good luck to most cricketers—unless it's your opponent's score.

How it works

To multiply a three-digit number by 111, let's call the number $100l + 10m + r$, where l, m, and r remind us of left, middle, and right. We form $(100 + 10 + 1) \times (100l + 10m + r)$. Calculating term by term gives $10,000l + 1,000(l + m) + 100(l + m + r) + 10(m + r) + r$. Thus the left digit gives the 10Ks, the sum of left and middle gives the thousands; all three sum to give the hundred. The middle and right give the tens, and the right digit gives the units.

Multiply or divide by 125

Multiplication

Numbers that end in 5 are gifts that keep on giving. To multiply a given number by 125 divide it by 8 and then add the number of zeros you need to get the right answer!

To multiply 126×125, estimate it as 100×100, so our answer should be about 10,000. Halve 126 to get 63, halve 63 to get 31.5, and halve 31.5 to get 15.75. The answer is $126 \times 125 = 15,750$.

As airlines sometimes say, usually after cramming you into a tiny seat and feeding you nothing but peanuts for a few hours, "We never forget you have a choice." Another choice you have to calculate 126×125 is to square 125. From the section "Square any number ending in 5," in Chapter 6 you can write down straight away that it is 15,625 and tack on the extra 125 to get 15,750.

Division

This is another easy one. Multiply your number by 8 (as in, double it three times) and add zeros as necessary. To find $1,263/125$, estimate it as $1,000/100 = 10$. Double 1,263 to get 2,526. Double 2,526 to get 5,052, and double 5,052 to get 10,104. Insert the decimal point to have the final answer $1,263/125 = 10.104$.

Multiply by $316\frac{2}{3}$, $633\frac{1}{3}$, and 950

There are probably few times in life you may be called upon to multiply any number by $316\frac{2}{3}$, but you can do so rather quickly and, arguably far more important, it gives insight into how to develop your own quicker-calculation techniques for your favorite numbers. The secret ingredient in this recipe is to average two numbers. Observe that if you have a number n, the average of $3n$ and $10n/3$ is:

$$\frac{1}{2}\left(3n + \frac{10n}{3}\right) = \frac{n}{2}\left(3 + \frac{10}{3}\right) = \frac{19n}{6} = 3.1666n$$

If you seek $316\frac{2}{3} \times 18$, triple the 18 to get 54. Divide the 18 by 3 to get 6 and affix 0 to get 60. Add them, which results in $54+60 = 114$, and halve to get the average, 57. The answer has to be about $300 \times 20 = 6,000$, and so we obtain our answer, $316\frac{2}{3} \times 18 = 5,700$.

To multiply by $633\frac{1}{3}$, follow the steps above but don't halve the result! That means $633\frac{1}{3} \times 18 = 11,400$. To multiply by 950, take the answer 5,700 and multiply by 3, which is 17,100.

Multiply by 999 or 1,001

To calculate any two-digit number by 999, multiply by 1,000 and then subtract the number. To multiply by 1,001, there's no need to: if you multiply by a three-digit number, simply write down the number twice. $234 \times 1,001 = 234,234$. If you multiply by a number of four digits or more, write down the number, shunt, and add.

For example, 999×23 *is* $23,000 - 23 = 22,977$. Multiplication by 999 can be done a slightly different way. In our example, subtract 1 from the 23 to get 22, and this is the 10K unit. Then do $1,000 - 23 = 977$ and bolt it on to the 22 to yield 22,977 as before.

And here we stop. 1,000 is, after all, the first time we can write down a number without using the letter "a." There are swift methods beyond this point (to multiply by 1,111, for example, you can use the swift method for 11, then shunt over twice and add) but by now I hope such a trick has become second nature to you.

Try these

1. 9.90×1.11
2. $4.5/125$
3. 11.1×0.72
4. 3.4×6.33333
5. 9.5×3.8
6. 0.29×99.9
7. 31.666×0.51
8. 6.6×111
9. 1.25×0.85
10. $9.1/12.5$
11. 3.1666×2.1
12. 7.8×63.33333
13. 6.1×0.111
14. 1.001×8.3

15. $7.7 \times 1,250$
16. 99.9×9.2
17. 1.11×28
18. 0.7×3.166666
19. $60/1.25$
20. $520 \times 11,100$
21. 3.5×0.95
22. $7.4/12.5$
23. $63/125$
24. 6.33333×7.8
25. 99.9×6.2
26. 1.2×0.125
27. 10.01×0.94
28. $4,300 \times 0.95$
29. 3.7×3.166666
30. 12.5×0.65
31. 11×125
32. $9,500 \times 1.3$
33. 2.20×3.166666
34. 100.10×0.31
35. 0.56×99.90
36. 2.90×3.166666
37. 9.50×0.65
38. 633.333×0.84
39. 1.40×316.6666
40. 3.10×9.99
41. 0.95×840
42. 633.333×9.60
43. 4.30×1.001
44. 41×633.333
45. $1.7 \times 9,990$
46. 5.60×10.01
47. $51 \times 1,001$
48. $9\frac{1}{2} \times 72$
49. 4.70×10.01
50. 87×6.333333

Interlude III

Doomsday

Write it on your heart that every day is the best day of the year. No man has learned anything rightly, until he knows that every day is Doomsday.

RALPH WALDO EMERSON, *Society and Solitude: Twelve Chapters*

"What day is it?" asked Pooh. "It's today," squeaked Piglet. "My favorite day," said Pooh.

A.A. MILNE, *Winnie the Pooh*

What day does Christmas fall on this year? The smug answer is "December 25, same as always," but that's the date, not the day. We know something else, too. That if July 1 is a Wednesday, then July 8 will be a Wednesday as well. In principle, if you know on which day of the week a certain date falls on in a particular year, you can work out what day of the week any other day will fall on that year. Another way to put this is to say each year has its Doomsday, mathematically speaking. British-born Princeton mathematician John Conway coined the term Doomsday, and using it allows those who can add or subtract quickly— those who have read this book so far—to be able swiftly to tell someone what day of the week a particular date falls on. The tricky part, beyond the scope of this book, is the actual calculation of the Doomsday for the year itself (if interested, see the Appendix!).

If you have a calendar for any year, past, present, or future, look up the day of the week on which April 4 falls. That day is the Doomsday for that particular year. Knowing this, we can fill in our mental calendar based on historic dates we've learned, and personal dates we will always remember.

Hints and tips help. In every year, April 4, June 6, August 8, October 10, and December 12 fall on Doomsday. That is to say, if Doomsday is Wednesday, then 4/4, 6/6, 8/8, 10/10, and 12/12 all fall on a Wednesday.

Immediately you know that D-Day, June 6, which marks the Allied invasion of Nazi-held Normandy during the Second World War, is always on Doomsday. By adding 14 days, we know that June 20 is also a Doomsday and therefore Juneteenth, June 19—which honors the emancipation of slaves in the United States—occurs the day before Doomsday. August 8 is Doomsday and so too is August 15, which marks Independence Day in India, which became independent of the United Kingdom on that date in 1947.

As 10/10 falls on Doomsday, if you add 3 weeks, or 21 days, you know that October 31, Hallowe'en, also falls on Doomsday, and November 1 (All Saints in some countries, and the Day of the Dead, Día de Muertos, in Mexico) is a day later.

Those from Mexico will also immediately recognize that 12/12 is the feast of Our Lady of Guadalupe—patroness of the Americas—marking the day in which she appeared to Cuauhtlatohuac, better known in the West as Juan Diego, who became the first indigenous saint from the Americas. And those in Great Britain and Canada can add a fortnight, to see that Boxing Day, December 26, is also on Doomsday.

A well-known chain of convenience stores in the United States is 7–11. The next phrase is worth learning: "Working 9 to 5 at 7–11." That's because May 9, September 5, July 11, and November 7 are all Doomsdays as well. Alas, in Europe and the United States dates are written differently, but no matter which side of the Atlantic you call home, 5/9, 9/5, 7/11, and 11/7 are all Doomsdays. Immediately, then, we know July 4, America's Independence Day, falls on Doomsday.

Keen readers will spot that there are no mnemonics presented so far for the first 3 months of the year. That's because leap years make things complicated. For March, that well-known date of March 0 is always on Doomsday, no matter what the year. March 0 is, as you might guess, the last day of February.

Important for those who love mathematics, we add 14 days to March 0 to surmise that March 14 is a Doomsday. This is π Day, (as written in the United States, it is 3/14) which was the birthday of the founder of the theory of relativity, Albert Einstein, the day that Stephen Hawking—who studied evaporating black holes and achieved popular fame with his book *A Brief History of Time*—passed away.

Two other phrases round out the picture. Either remember π Day, American style, as 3/14, or else remember the first three digits of pi, 3.14. That's because, when it is *not* a leap year, January 3, February *14*, and

3/14 are Doomsdays. February 14 is Saint Valentine's Day, a day set aside for celebrating loving relationships; ironically, it's the date on which Einstein's divorce from his first wife, Mileva Marić, became official.

If it *is* a leap year, try remembering a soccer score: January 4, February 1. Or, to give them names that sound like US soccer teams, January Leap 4, February Year 1. Both of these dates fall on Doomsday in a leap year. You can straight away find Doomsdays for important moments in your life. I broke my neck, playing rugby, on Wednesday November 14, 1984—a day and date that I will never have trouble remembering. From "Working 9 to 5 for 7–11," we know November 7 is Doomsday, and so, too, is November 14. Hence, Doomsday for 1984, a phrase George Orwell would have loved, was Wednesday. As April 4 is a Doomsday, therefore so is April 25, and we know that Shakespeare's birthday, usually marked on April 23, took place that year on Monday. And, as a fan of London Welsh rugby team, I'll point out that, as March 14 is Doomsday, so too is March 0, and so Saint David's Day fell on a Thursday that year.

With these few simple phrases committed to memory, you can impress people—especially married couples, such as parents and grandparents. Ask them to tell you the day and date when they tied the knot (often a Saturday). From that one date, you can compute the Doomsday for that year and rattle off various things, such as the day of the week of their first Christmas together, or on what day they first yelled "Ahoy maties!" in wedded bliss, on International Talk Like a Pirate Day (September 19, in case it slips your memory).

For example, celebrity couple Kim Kardashian and Kanye West married on Saturday, May 24, 2014. We know that May 9 is Doomsday, so by adding 14, we know May 23 was Doomsday. But, as May 24 was a Saturday, Doomsday for 2014 was Friday.

To add to your data bank, it's worth adding some day-related activities. For sports fans, the FA Cup Final is always held on a Saturday, while the Super Bowl takes place on a Sunday. Knowing that the so-called "White Horse Final"—which saw police horse Billie save the day and after whom Wembley's White Horse Bridge is named—took place on April 28, 1923, allows you to know that Doomsday for 1923 was Wednesday (for the record, Bolton beat West Ham United 2–0). The "Blackout Bowl," otherwise known as Super Bowl XLVII between the Baltimore Ravens and the San Francisco 49ers, played on February 3, 2013, which means that Doomsday for 2013 was Thursday. The Ravens ran out winners that night 34–31, though the 49ers staged a dramatic comeback after the lights came back on. And who could forget the last win by the United Kingdom in the Eurovision Song Contest, when Katrina and the Waves won in Dublin on Saturday May 3, 1997. But as May 9 is Doomsday, so is May 2, and therefore Doomsday was on a Friday that year.

There are other events keyed in to certain days. Americans might recall that the new session of the Supreme Court starts on the first Monday of October, and Canadian Thanksgiving is the second Monday in October. Elections in the United States are held on the first Tuesday in November, and Thanksgiving is always on a Thursday. And for Christians, Ash Wednesday is, well, on a Wednesday and Good Friday is, er, on a Friday.[1]

This gives you ample material, to which you can add more, with which to awe or bore your friends with calendrical factoids. If you know your friend's birthday, ask them to pick a year and tell you what date Canadian Thanksgiving fell, and seconds later, you can tell her on what day of the week her birthday fell, the day of her parent's anniversary, and other vitally important information. This trick will even save you money: you need never buy a calendar ever again—once you know the year's Doomsday.

---

[1] Ascension Thursday, for some odd reason, is celebrated on a Sunday.

5

Calculations with Constraints: Multiply and Divide by Numbers with Specific Properties

Bid all our sad divisions cease.

> From the last verse of John Mason Neal's
> rather free translation "O Come, O Come Emmanuel"
> of the Latin hymn "Veni, Veni Emmanuel"

Minus times minus equals plus
The reason for this, we need not discuss.

> W.H. AUDEN, A Certain World: A Commonplace Book

The previous chapter dealt with multiplying and dividing specific *numbers*. This chapter focuses on numbers that have specific *properties*. For some multiplications, therefore, techniques from both chapters might help you to work out the answer to the same problem in different ways. The delightful consequence of such richness is that you can choose whichever method is quickest. For example, when faced with 31×39, you could use the method for multiplying by 31 or 39 in the previous chapter, or the one in this chapter for multiplying two "kindred" numbers. The choice is yours, and the challenge is to find new methods of your own.

Multiply two numbers, 10 apart, ending in 5

If your great desire is to multiply 85×95, there's an easy way to do it. Pick the number exactly in the middle, in this case 90, and square it. That gives 8,100. Then subtract 25, to get 8,075.

By now, it's probably clear that the two numbers don't actually have to be 10 apart. For example, 6.5 and 7,500 are more than 10 apart, but we need to compute 65×75 and sprinkle in some zeros. Our estimate, rounding one number up and the other down, is that it's in the neighborhood of $6 \times 8,000 = 48,000$. Pick the number in the middle of 65 and 75, which is 70, square it to obtain 4,900 and subtract 25, which yields 4,875. Our answer, given the estimate, is $6.5 \times 7,500 = 48,750$.

The largest book in the United States' Library of Congress is John J. Audubon's *Birds of America*. First published as a series between 1827 and 1838, Audubon's book measures $2.5 \times 3.5'$.[1] To calculate the area of one of its pages, we know the answer is about 7 square feet, and spot immediately that 25×35 is the multiplication of two numbers, ending in a 5, separated by 10. We take the number in the middle, 30, square it, giving 900, and subtract 25, to get 8.75 square feet.

How it works

This relies on the difference of two squares. Write the two numbers differing by 10 *as* $(n-5)(n+5)$. By the difference of two squares, we know $(n-5)(n+5) = n^2 - 25$. The number in the middle is always a multiple of 10, and so is easy to square rapidly.

Multiply two numbers, 20 apart, ending in 5

Should you want to calculate 55×75, square the number in the middle, 65, then subtract 100, and you're done! As 65 ends in a 5, there is a quick way to do this (for more details, see the section, "Square any number ending in 5"!). It's 4,225. Subtract 100, to get $55 \times 75 = 4,125$.

Don't take "20 apart" literally. Should we want $4.5 \times 65,000$, the same approach holds. We can estimate that the answer is close to $4 \times 70,000 = 280,000$ and then ignore zeros and decimal points, as usual. The number halfway between 45 and 65 is 55, which, when squared, is 3,025. Subtracting 100 reduces this to 2,925, and so our answer is $4.5 \times 65,000 = 292,500$.

What, you may ask, does this have to do specifically with numbers ending in 5? You have made an astute point. This is a quick calculation

---

[1] For those not used to non-metric measures, $'$ marks a foot, which is 12 inches. Inches are marked as $''$. Spoiler alert: in the movie *This is Spinal Tap*, a heavy metal band seeks to recreate Stonehenge as a backdrop to their performance, but gets the signs for inches and feet confused. This gives you a plausible mathematical reason for watching the movie.

because squaring numbers ending in a 5 is blisteringly fast (see that section if you don't believe me!). But the method is more generally valid: when multiplying two numbers that differ by 20, square the number in the middle and subtract 100.

As proof, consider yet another example drawn from paper sizes. A folio of foolscap is $8\frac{1}{2}'' \times 13\frac{1}{2}''$ and is named after the jester's hat that used to be used as the watermark for such paper. At first glance, this has nothing to do with multiplying numbers that differ by 20. Rewrite it, though, as $\frac{17}{2} \times \frac{37}{2}$, and the game's afoot. Pick the number in the middle, 27, and square it, which is 729. Take off 100 to get 629. Now divide by 4, so we halve it to get 3,145 and do so again to get 15,725. Remember, we're saving time here by ignoring pesky decimal points. Sure, we could write that half of 629 is 314.5 and halving it again is 15,725, but we ignore these details because we'll put the decimal point in right at the end, to assure ourselves of the right answer. We'd estimate the area as about $9 \times 12 = 108$ square inches, so our answer is that a sheet of foolscap has an area of 157.25 square inches.

Yes, this does indeed take more time than entering the numbers into a calculator, but it's a great way to flex your calculating muscles by using such simple pre-calculator techniques.

And, with paper sizes fresh in the mind, let's round out this section by thinking of a typical size of paperback books in the United States, dimensions $5\frac{1}{2} \times 8\frac{1}{2}$ inches. Written differently, that's 5.5 × 8.5. Come up with a way to multiply rapidly two numbers, 30 apart, ending in a 5.

How it works

As with the previous method, this one works on the difference of two squares. We write the expression as $(n - 10)(n + 10) = n^2 - 100$.

Multiply a one- or two-digit number, less than 50, by 98

Immediately the mathematical nostrils sniff the faint scent of the number 100 lurking close by. And they are right to do so. An example shows the technique. If you want to calculate 47 × 98, subtract 1 from the 47 to get 46. Next, double the 47 to get 94 and take that from 100 to get 06. Bolt this on to the 46 to yield the answer:

$$47 \times 98 = 4,606$$

Why 06 and not just 6? Because we know that our answer is about 5,000 and so must have four digits. If we don't write the 0 in 06, we might wrongly assume our answer is 4,660.

Again, we can extend this to closely connected problems. For example, 0.036×9.8 is the same kind of sheep, albeit in wolf's clothing. Estimate this as roughly $0.04 \times 10 = 0.4$. Now compute 36×98. To do so, subtract 1 from 36 to get 35. Double the 36 to get 72, and when you subtract that 72 from 100 you get 28. Hence we get 3,528, which, given the estimate, means $0.036 \times 9.8 = 0.3528$.

Bananas are radioactive, courtesy of the Potassium-40 (^{40}K) they contain. A typical banana, when ingested, emits a dose of radiation equal to about 9.8×10^{-8} sieverts. A hand of bananas usually contains about 10 to 20 of the yellow fruit. If you eat 17 at one sitting, you would be exposed to 17 "banana equivalent doses" or BEDs, as they are known. This is $17 \times 9.8 \times 10^{-8} = 17 \times 98 \times 10^{-9}$. So, subtract 1 from the 17 to get 16. Double the 17 to get 34 and subtract that from 100 to obtain 66. Hence $17 \times 98 = 1,666$, and so scarfing 17 bananas gives you a radiation dose of 1.666×10^{-6}. sieverts.

Bananas aren't the worst, though. The average banana has about 9 disintegrations of ^{40}K per second (by emitting an electron, it decays into ^{40}Ca, but if it emits a positron and a gamma particle, it decays into ^{40}Ar). A mere 4 oz (113 g) of ground beef undergoes about 29 disintegrations, about the same as half a cup of red kidney beans. A bowl of chili, then, is one of the most radioactive main courses you can get. To set this in perspective, though, a chest CT scan in a hospital exposes you to an amount of radiation equivalent to eating 70,000 bananas.

The local acceleration owing to gravity, g, is about 9.8 m/s^2. Astronauts and pilots train to be able to withstand high g forces. But if the acceleration pushes you forward, (technically known as "eyeballs in" acceleration) you can withstand higher g forces, for longer, than an acceleration pushing you backwards ("eyeballs out"). In a NASA study from 1960, subjects could withstand $8g$ eyeballs in for about a minute, but about $7g$ eyeballs out.[2] You can now work out what $7g$ and $8g$ are in m/s^2. There can be other benefits of high g forces. The *Journal of the American Osteopathic Association* published an article that reported the case

[2] Brent Y. Creer (1960) "Centrifuge study of pilot tolerance to acceleration and the effects of acceleration on pilot performance." https://ntrs.nasa.gov/archive/nasa/casi.ntrs.nasa.gov/19980223621.pdf, p. 32 (accessed November 5, 2020).

of two doctors who built a kidney simulator, placed kidney stones in it, and then took it for 20 rides on Disneyworld's Big Thunder Mountain Railroad rollercoaster in Orlando, Florida. Four out of 24 kidney stones were passed due to the high-g rollercoaster motion.[3]

How it works

Call the two-digit number n. By definition, $98n = 100n - 2n$. The secret sauce is to write $98n = 100n - 100 + 100 - 2n$, and then recast it as $98n = 100 (n - 1) + (100 - 2n)$. Hence, you subtract 1 from n to get the hundreds and thousands digit, and take $2n$ from 100 to get the tens and units digits.

Multiply a one- or two-digit number by 99

This follows along the same path as multiplication by 98. An example shows the way.

Suppose we seek 23×99. First, subtract 1 from 23 to obtain 22. Next, form $100 - 23 = 77$. The answer is $23 \times 99 = 2,277$. Should you prefer, you could take the 23, glue two zeros on to the end to get 2,300 and then subtract 23, to get 2,277. The choice is yours.

How it works

This works almost identically to the previous section. Call the two-digit number n. Then by definition, $99n = 100n - n$. Again, note that $99n = 100n - 100 + 100 - n$, and recast as $99n = 100 (n - 1) + (100 - n)$. The hundreds and thousands column is therefore $n - 1$, while $100 - n$ gives the tens and units columns.

Multiply a one- or two-digit number by 101

The thing that is in Room 101 is the worst thing in the world.
GEORGE ORWELL, *1984*

This may not really count as a tip at all. To multiply by a single-digit number, write the number down twice with a zero in the middle: $7 \times 101 = 707$.

---

[3] Marc A. Mitchell and David D. Wartinger, "Validation of a functional pyelocalyceal renal model for the evaluation of renal calculi passage while riding a roller coaster," *The Journal of the American Osteopathic Association*, 116 (2016), 647–52. doi:10.7556/jaoa.2016.128

It's not much different when multiplying by a two-digit number. Write the two digit number down twice: $56 \times 101 = 5,656$.

But there is something mildly of interest about 101. It's a prime number, and the next prime number is 107. That's six more and, for reasons that are unclear, two primes, six apart, are called sexy primes.[4] If that's not enough to spark interest, observe that 113 is six more than 107, making 101 a member of a sexy triplet of primes.

Multiply a one- or two-digit number, less than 50, by 102

This is only a tad more difficult than multiplying by 101. For a single-digit number, write down the number, then write down double the number. So, $7 \times 102 = 714$. For a double-digit number less than 50, do the same thing: $23 \times 102 = 2,346$.

The "less than 50" requirement avoids having to worry about carrying a number over.

Try these

1. 45×55
2. 1.02×0.91
3. 98×47
4. 0.68×9.9
5. $3.6 \times 1,010$
6. 91×0.98
7. 1.1×9.9
8. 3.5×850
9. $780 \times 1,010$
10. 0.98×33
11. 0.45×750
12. $1.01 \times 9,500$
13. 98×0.51
14. 850×9.5
15. 990×0.038
16. 0.098×4.5
17. 0.25×0.35
18. 9.9×8.7

[4] The reason may well be that the Latin word for six is *sex*.

19. 850×1.05
20. $1.01. \times 1.7$
21. 95×11.5
22. 99×6.7
23. 5.5×750
24. 101×48
25. 350×5.5
26. 520×102
27. $0.85 \times 1,050$
28. 990×67
29. 7.2×1.02
30. 7.5×9.5
31. 210×101
32. 1.9×9.8
33. 102×0.26
34. 250×1.01
35. 4.3×980
36. 5.8×9.9
37. 10.2×0.62
38. 63×0.98
39. 0.31×0.98
40. 990×0.21
41. 8.8×99
42. 10.1×7.9
43. 8.5×0.99
44. 0.101×7.8
45. 620×10.2
46. 2.1×1.01
47. 41×10.2
48. 9.8×3.1
49. 102×450
50. 5.1×1.02

Multiply two numbers that differ by 2, 4, 6, or 20

This is a great trick to use if you can recognize when to use it—and if you know your squares! Take as an example, prime numbers. These are to numbers what atoms are to chemistry: the building blocks from which all other things are made. Unlike atoms, there is an infinite number of primes.[5] Like London buses or the DC Metro, you wait a while for a prime number to come along and then you get two in quick succession. Twin primes differ by two, such as 11 and 13, and there are mathematical theorems and conjectures about them. If you feel the urge to multiply twin primes together, then the following method works: find the number in the middle, square it, and subtract 1.

If you want to multiply 17×19, the number in the middle is 18, and:

$$18^2 = 324$$

Thus, $17 \times 19 = 323$. This works for all numbers, not just twin primes: $14 \times 16 = 15^2 - 1 = 224$.

If the numbers differ by 4, square the middle number and subtract 4.

$$19 \times 23 = 212 - 4 = 441 - 4 = 437$$

The game of Rugby Union, which Oscar Wilde said is "a good occasion for keeping thirty bullies far from the centre of the city," is played on a pitch whose length can be from $106-144$ m, and whose width is between 68 and 70 m (Rugby League is played on a field 68 m wide, but it is shorter, being only $112 - 122$ m in length). Consider a pitch of size 144×68 m. What's the area? Think of it as $2 \times 72 \times 68$. The last two numbers differ by 4, so we square 70 to get 4,900, subtract 4, to get 4,896, and double, to get 9,792 m^2, important to know if you need to replace the turf or install new underground heating pipes to prevent the pitch from freezing.

If the numbers differ by 6, square the middle number and subtract 9. Or, as you know by now, square the middle number, subtract 10, and add on 1.

$$13 \times 19 = 16^2 - 10 + 1 = 256 - 10 + 1 = 247$$

If the numbers differ by 20, square the middle number, then subtract 100.

$$14 \times 34 = 24^2 - 100 = 576 - 100 = 476$$

The important point here is that, as you've seen, there are many tips and tricks for numbers ending in 5, say. When you have a multiplication

[5] This is another example of a beautifully succinct proof by contradiction. If there were only a finite number of primes, you could multiply them all together and add 1. This new number isn't divisible by any of the allegedly finite primes, so it must be prime. Hence, there is an infinite number of primes. Try it! If the only primes were 2, 3, and 5, you could multiply and add 1, to get 31, which is prime.

that doesn't involve one of the easier tricks, it's precisely then that you should call to mind calculations like this.

The only downside is that you need to know your squares (there is a list at the end of this book, Appendix II). As parents have said for generations, "You've got to eat your vegetables." Think of learning your squares as eating your mathematical vegetables, a possibly unenjoyable process but with great benefits afterwards.

How it works

The difference of two squares is key. We have $(n - m)(n + m) = n^2 - m^2$. If the numbers differ by 2, then $m = 1$, and our answer is $n^2 - 1$. If they differ by 4, $m = 2$ and our answer is $n^2 - 4$, and so on.

Multiply two "kindred" numbers

This is another method to recall if any of the elementary ones don't work. Suppose you have two "kindred" numbers. By this, I mean two two-digit numbers that have the same first two digits—and this is important—whose last digits add up to 10. You can calculate these rapidly. First, take the tens digit, add 1 to it, and multiply them together. This will provide the thousands and hundreds digit of the answer. The tens and units column you get by multiplying the ones digits of the two numbers.

Consider 32×38. The tens digit is 3. Add 1 to get 4, and $3 \times 4 = 12$. Multiply 2×8 to get 16. The method says $32 \times 38 = 1,216$.

As a different example, ponder $4.3 \times 4,700$. As an estimate, this is about $4 \times 5,000 = 20,000$. With an estimate in hand, we remove all zeros and decimal points and concentrate on 43×47, a kindred-number multiplication. Add one to the 4 to get 5. Multiply the $4 \times 5 = 20$, which is the thousands and hundreds column. Next, multiply the $3 \times 7 = 21$, and write the two answers together to form 2,021. Given our estimate, we know $4.3 \times 4,700 = 20,210$.

How it works

The tens digit for both numbers is n. The unit digits are m and $10 - m$. The multiplication, then, is:

$$(10n + m)(10n + [10 - m]) = 100n^2 + 10n[10 - m] + 10nm + m(10 - m)$$

Simplifying:

$$(10n + m)(10n + [10 - m]) = 100n^2 + 100n + m(10 - m)$$

Or:

$$(10n + m)(10n + [10 - m]) = 100n(n + 1) + m(10 - m)$$

The first term is the hundreds digit, which is $n(n + 1)$, the result of adding 1 to the tens digit and multiplying by the tens digit. The tens and units digit of the product is $m(10 - m)$, which is the result of multiplying the two unit digits of the multiplicands.

Multiply by 23, 34, 45, or numbers "remainder 1" when divided by 11

By now, this kind of thinking should be almost natural. To multiply by 34, say, you don't have to worry about multiplying by 17 and doubling the answer. Instead, think of it as $33 + 1$. You then use the quick way to multiply by 11, multiply that by 3, and then add in the original number for the answer.

One of the suboxides of cesium is Cs_7O_2. The atomic mass of cesium is 133 amu, and for oxygen it's 16. The mass of 1 mole of this cesium suboxide is $(7 \times 133) + (2 \times 16)$. To show the method, this is $(7 \times 132) + 7 + 32 = (7 \times 12 \times 11) + 39$. First, do the multiplication $7 \times 12 = 84$. Use the rapid 11 method to get $84 \times 11 = 924 + 39 = 963$ grams/mol.

Those with little fear of subtraction can multiply by 21, 32, 43, and those numbers that are "remainder 10" when divided by 11, just subtract the original number at the end, rather than add it.

Multiply by 24, 36, 42, 48, or numbers where one digit is twice the other

The key point here is that because one number is double the other, you need only do one multiplication, double that number, and shunt to get the answer. If you seek 193×42, say, double 193 straight away to get 386. Double it again to obtain 772 and strap on a zero. Then add:

$$\begin{array}{r} 7,720 \\ +386 \\ \hline 8,106 \end{array}$$

This technique can be useful for large numbers if you go "number spotting." To multiply by 6,003, for example, you multiply by 3, double to get 6, and by shunting the numbers over and adding, you have a quick answer, just as if you have to multiply by 1.224.

Thomas Jefferson, the third president of the United States, wanted a different unit of time from the second we use today. After all, surely a new country deserves a new way to measure time. He designed a bar pendulum whose time to swing back and forth was 2 seconds. For this to happen, the length of the pendulum has to be 2.9806 m.[6] Twice this length, Jefferson decided, would be 10 Jeffersonian feet, so that 10 Jeffersonian feet are 5.96172 m. Each Jeffersonian foot would be 0.59172 m and, as there are 10 Jeffersonian inches (let's call them Jinches) in a Jeffersonian foot, each Jinch is 0.059172 m, or about 2.347 of the standard modern-day inch.

What does this have to do with doubling? Well, how long are 12 Jinches? Roughly speaking they must be 2 × 12 = 24 regular inches. The fastest way to get an approximate answer is to shunt. If you spot that 2.347 is almost 2.346 and note that 23 × 2 = 46, life becomes easy. We're ignoring decimal points, so think only of 2,346. To find our answer, 12 × 2.346, take four steps. First, multiply 12 × 23 and shunt twice, which is the same as multiplying by 2,300. Then we double our answer when multiplying by 23, which corresponds to multiplying by 46. Last, add them.

We have 12 × 23 = 276, which, when shunted left twice is 27,600. Double the original 276 to get 552 and add it on, to get 28,152. Bearing in mind our estimate of 24 inches, our answer must be 12 Jinches = 28.152 inches. In case it ever comes up on a quiz, the slave owner Thomas Jefferson was 6′ 2″, or 74 inches, tall. That's about 31.5 Jinches. Abraham Lincoln, who on September 22, 1862 signed the Emancipation Proclamation that ended slavery, and Lyndon Johnson, who signed the Civil Rights Act of 1964, are the tallest US presidents, who both measured 32.38 Jinches.

---

[6] See Alex Hebra, *Measure for Measure: The Story of Imperial, Metric, and Other Units* (Baltimore, MD: The Johns Hopkins University Press, 2003), p. 26. A bar pendulum of length L has a time period $T = 2\pi\sqrt{L/3g}$.

Multiply two numbers (or square a number) just under 100

It's easier, and faster, if both numbers are in the 90s.

Suppose we seek 93×98. The major ingredient here is not the 3 or the 8, but what you get when you subtract them from 10. In this case, 7 and 2. Multiply these together to get 14. These form the last two digits of the answer. Now you have a choice. You can subtract 7 from the 98, to get 91. Or you can subtract the 2 from the 93, which gives the same answer. But 91 forms the first two digits of the answer. We have, then, $93 \times 98 = 9,114$.

This allows you to calculate the squares of numbers in their 90s speedily, as they are merely special cases of the above.

To calculate 97^2, take $10 - 7 = 3$, square it, so the last two digits are 09. Now subtract 3 from 97 to get 94, so our answer is $97^2 = 9,409$.

As so often before, don't get too literal in hunting for such multiplications. To compute 9.2×9.6, use the same method. The answer is between 81 and 100. Focus, then, on 92×96. Do the subtraction $100 - 92 = 8$, and also $100 - 96 = 4$. Multiply these together to get $8 \times 4 = 32$, which are the last two digits of the answer. We now need to take the 8 away from the 96 (or a 4 away from 92), which gives 88. Glue on the 32 to get 8,832 and, recalling the estimate, insert a decimal point to get $9.2 \times 9.6 = 88.32$.

How it works

When multiplying two numbers in their 90s, write them as $(90 + n)$ $(90+m)$. Transform this into $[100-(10-n)][100-(10-m)]$. The first step, then, calculates $10 - n \left(= a, \text{say} \right)$. and $10 - m \left(= b, \text{say} \right)$. We are therefore calculating $(100 - a)(100 - b)$. Multiplying term by term gives $100(100 - a - b) + ab$. The tens and units of the answer equals $a \times b$. The hundreds are determined by $(100 - a - b)$. But $(100 - a - b) = (90 + n) - b$, which is one of the original numbers from which "10 − the other number" has been subtracted.

Multiply two numbers (or square a number) just over 100

As the two numbers are "just over" 100, the answer will be a five-digit number. The first of these will be 1. If one number is $(100 + n)$ and the other is $(100 + m)$, then the hundreds column equals $n + m$, and the final two digits are given by $n \times m$.

To multiply 104×107, the sum of the ones digits is $4 + 7 = 11$, while their product is $4 \times 7 = 28$. Thus $104 \times 107 = 11, 128$.

The same approach works (see "How it works" below for the warning) for numbers in the $110 - 19$ range. Consider 113×107. We have $n = 13, m = 7$. Adding them gives 20. Multiplying gives 91, so we write down $113 \times 107 = 12, 091$.

This technique allows for a stylish application: calculating the squares of the numbers from 101 to 109 in a heartbeat. Ask your friends, students, or audience for a number in this range and you can shout out its square in a second or so: $108 \times 108 = 11, 664$.

The number 108, by the way, is a so-called powerful number. If there is a prime number p that divides 108 so, too, does p^2. The numbers 3 and 9 divide 108, as do 2 and 4. Mathematician Sol Golomb ($1932 - 2016$) coined the phrase "powerful numbers," and his creation of mathematical objects known as polyominoes serves as the bases of the popular computer game Tetris.

How it works

The numbers are $(100 + n)(100 + m) = 10, 000 + 100(n + m) + nm$. In this context, "Just over" becomes clear. Namely, we need the product $n \times m$ to be less than 100. When that holds true, the answer begins with a 1 in the 10K column, has a hundreds (and thousands) column determined by $(n + m)$, and a units column nm, with no need to worry about carrying numbers over.

Multiply two numbers, either side of 100

We can combine the previous two sections. Suppose you wish to multiply 93×108. First, you will have 10 in the thousands column. 93 is 7 short of 100, so the next step is to calculate 8–7, which is 1, and this is the hundreds column. So far, then, we have 10,100. The last step is to multiply the 7 and 8 to get 56, subtract this from 10,100 to obtain $93 \times 106 = 10, 044$.

For 97×102, we write down 10,000. The hundreds column is now $2 - 3$, and so we subtract 100 to get 9,900. Now we subtract 3×2 to get $97 \times 102 = 9, 894$.

How it works

We are calculating $(100 - n) \times (100 + m) = 10,000 + 100(m - n) - nm$.

Multiply together two two-digit numbers

The key piece of information here is that you can do this without showing any work, something guaranteed to impress fellow mathematics enthusiasts. The technique is simple. Multiply together the ones column, which gives the ones column of the answer. Cross multiply the tens column of one with the ones column of the other, and vice versa, and add these. This gives the tens column. Then multiply the tens columns together to get the hundreds column.

As an example, suppose we seek 76×84. Multiply the ones columns to get 24, and so we can write down a 4 in the ones column. Next we form 7×4 (which is 28) add to 6×8 (which is 48) to get 76, and we add the 2 from the previous step. We thus have 78 and can write down the middle digit as 8. Last, we multiply 7×8 to obtain 56 and add the 7 to get 63. Hence we have 6,384. But, going back to basics, remember our goal is to speed up but to avoid mistakes while doing so. It's probably better, then, to form $7 \times 4 = 28$ and add in the 2 that's been carried right away, just so you don't forget it, to get 30. Then add the $6 \times 8 = 48$ to get 78.

How it works

This is what it means to multiply! We are taking two numbers, which we can write as $10n + m$ and $10u + v$ and multiply them together. This is $(10n + m) \times (10u + v) = 100(n \times m) + 10(n \times v + m \times u) + m \times v$. If we don't have to carry any numbers over, then this means the units column is what we get by multiplying the two units together; the tens column we get by cross multiplying the digits in the numbers, while the hundreds column is the product of the tens columns.

Try these

1. 55×57
2. 7.1×790
3. 1.8×23
4. 6.3×6.5

5. 0.81×83

6. $1,010 \times 1.09$

7. 3.40×0.32

8. 770×8.1

9. 1.3×1.7

10. 280×320

11. 0.61×0.65

12. 8.2×860

13. 6.30×0.69

14. 640×7.8

15. 7.1×7.7

16. 5.9×56

17. 10.3×102

18. 310×2.1

19. 53×59

20. 0.9600×0.0092

21. 13×2.3

22. 7.1×9.1

23. 9.4×11.4

24. 13×1.8

25. 2.2×440

26. 9.7×970

27. 9.9×9.2

28. 9.5×970

29. 9.1×9.4

30. 1.03×1.05

31. 1.04×9.90

32. 1.05×930

33. 1.07×9.30

34. 1.04×98

35. $1,080 \times 980$

36. 10.3×9.9

37. 7.6×8.2

38. 36×96

39. 140×710

40. 1.20×0.36

41. 38×4.5

42. 3.2×210

43. 1.50×0.57

44. 3.2 × 2.7
45. 1.7 × 130
46. 24 × 26
47. 32 × 38
48. 55 × 55
49. 4.80 × 0.42
50. 330 × 8.7

Interlude IV

Multicultural Multiplication

There are nine and sixty ways of constructing tribal lays,
And every single one of them is right!
RUDYARD KIPLING, "In the Neolithic Age"

There are 10 types of people in the world: those who understand
binary and those who don't.
Popular mathematics T-shirt slogan

Latin, the language of Ancient Rome, greatly influenced many lan-
guages in the West. Imagine, though, if the language of mathematics
had remained Roman. This book is to help readers calculate more
quickly. Pause for a moment: how quickly can you calculate using
Roman numerals. Take an earlier section, for instance, multiplying
two numbers just over 100. The example was 104 × 107. With Roman
numerals, the section would be called "Multiply two numbers, just over
c," and you would need to compute civ × cvii. Indo-Arabic numerals
are far easier. The story of how they spread into the West is fascinating. Early advocates included Gerbert d'Aurillac (946–1003), who
later reigned as Pope Sylvester II and who remains the only Pope to
have his mathematical works published—they take up four volumes!
The use of Indo-Arabic numerals spread slowly, though, a sure sign of
how resistant we can be to change. Bankers who worked in Florence,
Italy, were still required to use Roman numerals as late as 1299. The
University of Padua insisted that book prices be marked not in figures,
but in "clear letters."[1]

---

[1] "*Non per cifras, sed per literas claras.*" As quoted by Isaac Taylor, *The Alphabet: An
Account of the Origin and Development of Letters, Vol. II* (New York: Scribner's Sons, 1899),
p. 263. Taylor says that Indo-Arabic numerals were first adopted in the West for use in
mathematics, then for page numbers in books, and took until the fifteenth century to
become widespread. Even today the so-called front matter of books (title pages, tables of
content, prefaces, and the like) bears page numbers in Roman numerals, while the main

Indo-Arabic numerals are far from being the only contribution to mathematics from non-Western cultures. A variety of cultures—such as those of Babylon, India, Egypt, and China among others—made great strides in mathematics. Naturally, a short book such as this cannot do them justice. Instead, let us look at a small number of methods that permit some quick calculations, ones that perhaps aren't at all familiar from classroom or textbooks.

Indian multiplication

The history of mathematics can be difficult to untangle. Good ideas tend to be shared and improved upon, so tracing back to find out who discovered or invented what, and who learned or borrowed from whom, is fraught with difficulties.

There are a number of ways in non-European cultures to multiply. We want to focus on relatively quick methods. The Indian mathematician Bhāskara II (1114–1185) in his work *Līlāvatī* (the name of one of his daughters) outlined the grating or lattice method, as it is known in the West. The story goes that as Līlāvatī wasn't going to get married, her father gave her a book of mathematics to make up for it.

The lattice method begins with drawing, not surprisingly, a lattice. To multiply 34 × 72, say, begin by drawing a box and labeling it.

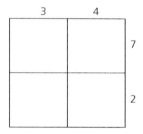

Figure 1 The basic lattice for 34 × 72

The second step, which isn't really a step, is to add diagonals to the boxes and extend them beyond the edges of the initial boxes.

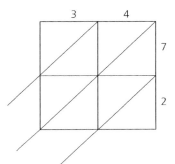

Figure 2 Extending the lattice

Focus on the top right square. It has a 4 above it, and a 7 to the right. Multiplying these two together gives 28. The 2 is written above that square's diagonal, the 8 below it. This is one of the four partial products for the lattice. Fill these numbers in for the entire square. The result is:

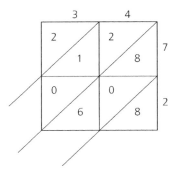

Figure 3 Calculating the partial products

The next step is to look at the diagonal lines. Sum the number contained between the two lines. Add also the sum of the numbers below and above both sets of lines. This gives (see the bold):

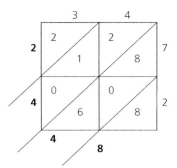

Figure 4 Summing along the diagonals

Notice that as $6 + 8 = 14$, we report only the 4, but add the 1 to the next sum. That means there is an important order in doing this: you go from the bottom right (below the diagonals) to the top left (above both diagonals). Bhāskara's method leads to the result $34 \times 72 = 2,448$.

For multiplication of two two-digit numbers, this may not be that impressive. But the same method will work, and in short order, for more complicated multiplications, such as 12.34×789. Try it!

The lattice method works by splitting up, in graphical form, the 34 into $(30 + 4)$ and the 72 into $(70 + 2)$.

Egyptian multiplication

Mathematics taught in school centers squarely on base 10. Those interested in computer science may learn something about base 2 (binary) or base 16 (hexadecimal, or hex for short). There is usually some moment when, in dealing with exponents, students learn how to convert from one base to another.

In Ancient Egypt, though, there was a method for swift multiplication based on binary.

Suppose we seek 225×17. Pick one of these two numbers, say 225. Write out two columns, the first one (on the left) headed by the number 1, the second (on the right) headed by 225. Double both columns until the left-hand column is about to exceed the multiplicand, which in this case is 17. The result is this:

1	225
2	450
4	900
8	1,800
16	**3,600**

Look down the left-hand side, and recall that $17 = 16 + 1$ (which is why the 16 and 1 are in bold!). Inspecting the right-hand side, the corresponding entries are 225 and 3,600. The Egyptian method says:

$$225 \times 17 = 225 + 3,600 = 3,825$$

How it works

The left-hand column represents powers of 2. By adding up those powers of two that make the multiplicand, we express the multiplicand as a (unique) number expressed as powers of 2. In this case we say $17 = 2^0 + 2^4$, so that $225 \times 17 = \left(225 \times 2^0\right) + \left(225 \times 2^4\right)$. This, though, is the sum of the number 225 itself, plus the number 225 doubled four times, but that's what we have already done in the right-hand column—all that doubling—and so gives us exactly what we need. If we wanted 225×22, then as $22 = 16 + 4 + 2$, we would add $3,600 + 900 + 450$ (to do this quickly, add 1,000 to the 3,600 to get 4,600 and add on 350 $[= 450 - 100]$). The answer is $225 \times 22 = 4,950$.

"Russian peasant" math

This unfortunate phrase was coined in the book *The Crest of the Peacock.* The technique closely resembles the Ancient Egyptian method. Write out two columns and, for the same problem as we had before, label one 225, label the other 17. Halve one and double the other repeatedly. When halving, ignore all remainders (round down). At the end, add up those numbers where the "halved" column is odd.

225	17
112	34
56	68
28	136
14	272
7	**544**
3	1,088
1	**2,176**

The odd rows on the halved column are in bold. The answer, according to this method, is $225 \times 17 = 17 + 544 + 1,088 + 2,176 = 3,825$. The secret here is, once again, binary digits. When you divide by 2, the remainder in binary is either 0 (if the number is even) or 1 (if the number is odd). Take a look at the steps involved. With step 0, we have 225. Step 1, after we have divided by 2, we have 112. Step 2, we have 56, step 3 has 28, step 4 results in 14, step 5 in 7, step 6 in 3 and step 7, after division by 2^7, we have 1. This means that when we divide 225 by 2^7, the remainder is 1. Likewise, the remainder is 1 when divided by 2^6 and 2^5 and by 2^0. Another way of viewing this is to say that we can express 225 in powers of two, and thus in binary, as $225 = 2^7 + 2^6 + 2^5 + 2^0$. Therefore $225 \times 17 = \left(2^7 + 2^6 + 2^5 + 2^0\right) \times 17$. This is exactly what we get, though, by looking at the numbers in the right-hand column, where we began with 17 (or, equivalently, $2^0 \times 17$). After doubling 17 seven times, we have $2^7 \times 17 = 128 \times 17 = 2,176$. and so on. Even though explaining the basis for the method takes a while, carrying it out is relatively swift. Using Egyptian numerals, however, would probably slow us down; we have the luxury of using Indo-Arabic numerals.

6

Super Powers: Calculate Squares, Square Roots, Cube Roots, and More

Power tends to corrupt and absolute power corrupts absolutely.

JOHN EMERICH EDWARD DALBERG-ACTON, Lord Acton,
Letter to Mary Gladstone, April 24, 1881

For the love of money is the root of all evil.

1 Timothy 6:10 (King James Version)

The square root of naff all.

Idiomatic English expression

I'm very well acquainted, too, with matters mathematical, I understand equations, both the simple and quadratical, About binomial theorem I'm teeming with a lot o' news, With many cheerful facts about the square of the hypotenuse.

"The Major-General's Song" from Gilbert
and Sullivan's *Pirates of Penzance.*

Mathematics deals with equations, and at the heart of an equation lies, perhaps obviously, the equals sign. The two parallel lines, $=$, an integral part of every computer keyboard or electronic calculator, are a comparatively recent invention, usually credited to a little-known Welsh mathematician by the name of Robert Recorde. He was born in Tenby, Pembrokeshire, in 1510, and was a great innovator in mathematical notation. Not only did he replace the Latin abbreviation "aeq." with his sign ("because no two things can be more alike") in his book *The Whetstone of Witte* (published in 1557), he also popularized the use of the $+$ sign among English-reading mathematicians. Not everything he did was successful. In the same book that gave the world the equals sign, he also introduced the word zenzizenzizenzic. The zenzizenzizenzic of a number n is a long way of saying n^8. It never caught on. But, while there is no swift way to find the eighth power or root of a number, we should recall that $8 = 5 + 3$. And it turns out that there are quick ways to find the fifth root, and the cube root, of a number, as this chapter will show.

To calculate quickly, it helps to know some squares. Knowing the squares of numbers from 1 to 30 can be beneficial to your health. For example, physicians believe that the body mass index (BMI) is a measure of overall health risks, and BMI is your mass, in kilograms, divided by the square of your height, in meters. Belgian astronomer and mathematician Adolphe Quetelet developed the formula in the mid-1800s, as a measure of good health. Let's apply it to actor Joe Pesci (of *Home Alone*, *Goodfellas*, and *My Cousin Vinnie* fame). He stands 5′ 3″ tall, which is 1.6 m. He weighs, as per the internet, some 58 kg. By knowing that $16^2 = 256$, we can guess his BMI is 58/2.56. As we shouldn't believe everything we read on the web, we can make life easier by using a denominator of 250, which after rapid division by 25 gives a BMI for Pesci of 23.2, which is in the healthy range for weight.

Joe Pesci probably would not want to have had to go several (or any) rounds in the boxing ring with Muhammed Ali, even if "The Greatest" was famously decked by Our 'Enry, Britain's Sir Henry Cooper. Ali was 1.9 m tall and weighed in at 107 kg (technically, he massed in at 107 kg, as kg is mass, while Newtons measure weight). Ali's BMI is then $107/(1.9^2)$. If you know your squares, this is 107/3.61, and the eagle-eyed will estimate this as roughly 108/3.6, which is 30. According to the US Department of Health, this would make Muhammed Ali obese, though I, for one, would not have dared to say that to him. What the numbers do show, though, is that the BMI formula has issues accounting for those with large amounts of muscle.

To whet the mathematical appetite, think of a four-digit number. I choose 1,458, as 1,458 is the year of the founding of Magdalen College, Oxford, alma mater of C.S. Lewis, Oscar Wilde, and Cardinal Wolsey, among others. Take the digits one at a time, square them, then add. $1^2 + 4^2 + 5^2 + 8^2 = 1 + 16 + 25 + 64 = 106$ (and the way to have done that quickly is "mystically," by doing $64 + 16$ first to get 80). Now repeat, to get $1^2 + 0^2 + 6^2 = 37$. And again, to get $9 + 49 = 58$. The next few stops on this tour are 89, 145, 42, 20, 4, 16, 37, 58. . . Do 37 and 58 seem familiar? We have stumbled on the so-called Steinhaus Cyclus. No matter which four-digit number you start with, this process will lead either to a 1, or else to the cycle 42, 20, 4, 16, 37, 58, 89, 145.

Benjamin Banneker published the first of his almanacs in 1792, which contained the data he calculated for lunar cycles, planetary conjunctions, and eclipses. This would be the first of several such works he compiled, and whose astronomical data he computed. Banneker, an African American, corresponded with Thomas Jefferson, sending him a copy of the Almanac, but in addition castigated the signer of the Declaration of Independence for hypocrisy regarding slavery.[1] Choosing 1,792 as our four-digit number, we calculate the squares of the digits and add them to get $1 + 49 + 81 + 4 = 135$. The next iteration is $1 + 9 + 25 = 35$. Then $9 + 25 = 34$, leading to $9 + 16 = 25$, and then to $4 + 25 = 29$. We are almost home: $4 + 81 = 85$, and $64 + 25 = 89$, and we have reached one of the numbers in the Steinhaus Cyclus.

Calculating the Steinhaus Cyclus only requires the ability to square single-digit numbers and add them. But what if you want to square two- or three-digit numbers? There are some short-cut methods to compute them.

Square any two-digit number ending in 1 or 9

In *The Mathematical Classic of Master Sun*, which dates to 400 BCE, the reader is set the problem of squaring 81.[2] To do so, write down the square of the number immediately below it, $80^2 = 6,400$. Double the 80 to get 160, and add on the 1. Thus, $81^2 = 80^2 + 80 + 80 + 1 = 6,561$.

In the Bible, in the book of Exodus (25:10), the Hebrew people are asked to build the Ark of the Covenant out of acacia wood, of dimensions $2\frac{1}{2}$ cubits long, $1\frac{1}{2}$ cubits wide, and $1\frac{1}{2}$ cubits high. In the Imperial system of units, the Ark is $52 \times 31 \times 31$ inches. The volume is therefore 52×31^2 cubic inches. But $31^2 = 30^2 + 60 + 1 = 961$. The volume of the Ark of the Covenant is 52×961 cubic inches, which is $(50 \times 961) + (2 \times 961) = 48,050 + 1,922 = 49,972$ cubic inches or, to use the measurements in Exodus, 5.625 cubic cubits.

---

[1] Banneker wrote that Jefferson was "detaining by fraud and violence so numerous a part of my brethren under groaning captivity and cruel oppression, that you should at the same time be found guilty of that most criminal act, which you professedly detested in others, with respect to your Selves."

[2] The Chinese method to calculate this is explained in Roger Hart, *The Chinese Roots of Linear Algebra* (Baltimore, MD: The Johns Hopkins University Press, 2011), pp. 14, 15.

To square a number ending in 9, do the opposite. If you wish to square 59, write down the square immediately above it, $60^2 = 3,600$. Subtract from this double 60, and then add 1. So $59^2 = 60^2 - 60 - 60 + 1 = 3,600$ $-120 + 1 = 3,481$. (Depending on how you feel, you could subtract 100 from 3,600 to get 3,500 and then take away a further 20, giving 3,480, and add on the remaining 1. Or you could take 200 from the 3,600 to get 3,400, and then add 80 to get to 3,480. Adding the 1 to get 3,481.)

There are some curiosities here. If you square 21, that's $400 + 40 + 1 = 441$. So $12^2 = 144$ and $21^2 = 441$, all you have to do is write backward! The very next number works exactly the same way. $13^2 = 169$. while $31^2 = 961$. Those are two easy squares to add to your memory's data bank.

How it works

For a square, $(n + m)^2 = n^2 + 2mn + m^2$. If $m = 1$, we have:

$$(n + 1)^2 = n^2 + 2n + 1$$

The method works quickly simply because n is a multiple of 10, which allows us to square it easily.

If we choose $m = -1$, we get:

$$(n - 1)^2 = n^2 - 2n + 1$$

Which is what we're using for the numbers ending in 9. Again, it works quickly because, by design, n is a multiple of 10.

Square any number ending in 5

This is a rather neat trick. There is a number in front of the 5. Add 1 to that number and then multiply those two numbers together. Write them down and append the digits 25 to the end. That's the answer.

For 85^2, take the number in front of the 5, which is an 8. Add 1 to get 9. Calculate $9 \times 8 = 72$ and then strap 25 to the end to get $85^2 = 7,225$.

Sometimes such calculations are slightly hidden. The maximum length of a Gaelic football pitch measures 145 m, with a maximum width of 90 m. The largest possible playing surface is thus 145×90. If the entire pitch needs to be returfed at the end of the season, how much will you need? This can be recast, explaining every stage, as:

$$145 \times 90 = (100 \times 90) + (45 \times 90) = 9,000 + \left(2 \times 45^2\right)$$

Using the trick for swift squaring of two-digit numbers ending in a 5, we can write this as:

$$145 \times 90 = 9,000 + (2 \times 2,025) = 13,050 \text{ m}^2$$

Naturally, you may prefer simply to do $145 \times 90 = 14,500 - 1,450 = 13,050$, but having alternative routes to get to the same answer is a hallmark of rapid math.

Please also remember to look for further applications. For example, 38×35 is not a square, but viewed as $(35 + 3) \times 35$, you can write it down straight away as $1,225 + 105 = 1,330$.

How it works

We know:

$$(a + b)^2 = a^2 + 2ab + b^2$$

We are squaring the number $10n + 5$. Plugging this into the formula gives:

$$(10n + 5)^2 = 100n^2 + 100n + 25$$

Rewriting:

$$(10n + 5)^2 = 100n(n + 1) + 25$$

This is the basis for the method. The last two digits will always be 25. To get the hundreds digit, add one to the tens digit, n, and multiply to get $n(n + 1)$.

Square any number ending in 6 or 4

The reward for learning the simple trick of squaring any number ending in 5 is to be able to square easily any number ending in 6. An example shows the method. Suppose we seek 46^2. Subtract 1 from 46 to get 45. Square the 45, which you can do instantly, as it is 2,025. Double the 45 to get 90 and add this in to get 2,115. Last, add a 1 to end up with 2,116.

In days gone by, the area of the country estate of the local aristocracy was measured in acres. An acre, to make things less clear, is an area that is 1 chain in length and 1 furlong wide. A chain is 22 yards, or 66 feet. A furlong is 10 chains, which is 660 feet. How many square feet are there in an acre? It's $66 \times 660 = 10 \times 66^2$. To find 66^2, we square 65 to get 4,225, then double 65 to obtain 130 and add it on to get 4,355. Now include the extra 1, and don't forget to multiply the entire answer by 10, to predict that an acre is 43,560 square feet.

Adapting this slightly, you can square any number ending in 4. It's the same process, except you subtract twice the number instead of adding it. For example, 84^2 is calculated by taking 85^2 $(= 7,225)$, subtracting 170 (to get 7,055), and adding 1 to get $84^2 = 7,056$.

What is 1 square inch expressed as square centimeters? We seek 2.54^2. First, square 255 to get 65,025. (If 25×26 seems daunting, remember the method for multiplying by 25: divide by 4. Halve the 26 to get 13; halve again to get 65. OK, when you halve 13 you should have 6.5, but we're ignoring decimal points, relying on our estimate to cough up the correct answer at the end. As the answer must be about $25^2 = 625$, this gives us 650, to which we append the usual 25 to get 65,025.)

Double 255 to get 510. Subtract this from the 65,025 to get 64,515 (as usual, take off 500, then take off 10 more). Then add one to get 64,516. As our answer must be about 4, our final answer is 1 square inch is 6.4516 square centimeters.

How it works

This technique uses $(n + 1)^2 = n^2 + 2n + 1$. As n in this case is a number ending in 5, we can square it rapidly; $2n$ is then a number ending in 0, which are usually easy to compute mentally and stir in. Including the final 1 is straightforward.

Square any two-digit number

Suppose we wish to square 57. Look for the closest multiple of 10, which is 60. Do this because it is far easier to multiply by a number that ends in a zero! Now, $60 - 57 = 3$. The next step, then, is to subtract that 3 from 57 to get 54, and then form $60 \times 54 = 3,240$. The last stage is to add on $3^2 = 9$. to get $57^2 = 3,249$.

Presented with the challenge of calculating 7.8×79, one way to do so would be to estimate it as $8 \times 80 = 640$. Then focus on $78 \times 79 = 78^2 + 78 = 78^2 + 80 - 2$.

To square 78, we go to the nearest 10, which is 80. Form $80 - 78 = 2$. Subtract 2 from 78, to get 76, then compute $80 \times 76 = 6,080$, to which we add the 2^2 to get 6,084. Adding 80 gives 6,164 and subtracting 2 results in 6,162. Based on the initial estimate, $7.8 \times 79 = 616.2$.

For numbers slightly more than a multiple of 10, simply reconfigure this. To square 52, say, we'd go for 50, and add the 2 to the 52 to get 54. We then form $50(54) = 2,700$ and add on 2^2 to get $52^2 = 2,704$.

How it works

Write the number you wish to square as $(n + m)$. Then:

$$(n + m)^2 = n^2 + 2nm + m^2 = n(n + 2m) + m^2$$

The first step in the calculation gives $n(n + 2m)$, the second is to add on m^2. Naturally, it's easiest to select n to be a certain multiple of 10 and choose m to be the smallest value possible. In the example, we could have used $57 = 50 + 7$, then formed $50(64) + 7^2 = 3,249$, but it's quicker to use 60 as it's closer to 57, as working with smaller numbers is usually faster and safer.

Square any three-digit number with middle digit 0 or 9

To calculate $(403)^2$, focus on the 4 and 3. Square the 4 to get 16. Then square the second number—making sure it has two digits. As 3^2 is 9, this is 09 (the two-digit requirement means that you should append 01, 04, and 09 for 1^2, 2^2, and 3^2, respectively). Multiply the 4 and the 3 together to get 12, and double to get 24. The answer we seek is $(403)^2 = 162,409$. The answer to $(809)^2$? Start off with 64, end with 81, but now you have 144 when you multiply the digits together and double them. We need to carry one over, then, to get 654,481.

If the middle digit is a 9, the only difference is to subtract rather than add—and to cheat a little! To find $(293)^2$, don't think of the first number as a 2, but as a 3, as we're trying to find $(300 - 7)^2$. Square the first number, square the second, and add. That's 90,049. Now subtract $2 \times 7 \times 300 = 4,200$. So $(293)^2 = 90,049 - 4,200 = 90,049 - 4,000 - 200 = 85,849$.

How it works

Write the number in the form $(100n + m)$. We know that $(100n + m)^2 = 10,000n^2 + 200nm + m^2$. By squaring the first digit, then, we have the 10K column of our answer. Squaring the second number gives the units column. The hundreds column is given by multiplying the two digits together, doubling, and affixing two zeros. For 409^2 the first digit is a 4, which, when squared, is 16, and so corresponds to 160,000 in our answer. The second digit, when squared, is 81. Multiplying them together gives $4 \times 9 = 36$, doubling gives 72, and gluing on two zeros gives 7,200. The answer is thus 167,281.

Square any three-digit number with middle digit 4 or 5

This follows the previous section closely, the only wrinkle being that you need to recall how to square a two-digit number ending in 5 (see that section!).

The title of Ray Bradbury's famous book *Fahrenheit 451* stems from the fact that paper burns at that temperature. (For a challenge, calculate 451 in centigrade.) For fun, let's calculate 451^2. Begin by computing the square of 45, which is 2,025. Then square the 1, but write it in a double-digit form, 01, and strap on to the previous number to form 202,501. Last, double the 45 to get 90, and multiply by 1, to get 90. Append a zero to form 900. Add $202,501 + 900 = 203,401$.

There is a practical application, one we're familiar with by now. Namely, 1 inch is approximately 2.54 cm. One square inch corresponds to $(2.54)^2$ cm^2. Hence, to convert from square inches to square centimeters, we need to know how to compute 2.54^2, for which we can use the technique above. A simple estimate is to say 1 square inch is about $2 \times 2 = 4$ cm^2.

The square of 25 is 625, to which we append the 16 that comes from squaring the 4, which generates 62,516. Double the 25 to obtain 50 and multiply by 4 to get 200. Appending a zero produces 2,000. Add $62,516 + 2,000 = 64,516$. Last, insert the decimal point at the right place, which yields 1 square inch $= 6.4516$ cm^2.

If the middle number is a 4, not a 5, you subtract the second number, rather than add it. To square 247, write it as $(250 - 3)^2$. The answer is $62,509 - (25 \times 2 \times 3 \times 10) = 62,509 - 1,500 = 61,009$.

Squaring, sort of. . .

For rest and relaxation, how would you work out π^2? As π is an irrational number, its decimal expansion never ends. To come up with a swift estimate, replace $\pi = 3.141592654$ by its Archimedean approximation, $\pi = \frac{22}{7} = 3\frac{1}{7}$. The rough answer is:

$$\pi^2 \approx \left(3 + \frac{1}{7}\right)^2$$

As 1/7 is small compared with 3, replace the square on the right-hand side by adding the square of 3 to twice the product of the terms in the bracket. This gives:

$$\pi^2 \approx 9 + \frac{2 \times 3}{7} = 9\frac{6}{7}$$

Recall the decimal expansions for fractions involving 7 to get:

$$\pi^2 \approx 9.\overline{857142}$$

There is an obvious objection that there are large numbers of places quoted after the decimal point, whereas the calculation neglects the term 1/49, or roughly 0.02. So the answer is "about 9.8." The good news is that whenever you see a calculation involving π^2, you can use the method for rapid multiplication by 98.

As another way of doing the same thing, we could choose to write:

$$\pi^2 \approx \frac{1}{100}(31 + 0.4)^2 \approx \frac{1}{100}\left(31^2 + 24.8\right) = 9.858$$

The quicker way to compute 31^2 is contained in the section "Square any two-digit number ending in 1 or 9," and the process is to square, double, and add 1 to the number immediately below. Thus $31^2 = 30^2 + 60 + 1 = 961$.

The method, taken as a whole, works best if there's a small digit involved somewhere.

As another example of an irrational number, consider $e = 2.71828...$ To square it, we'd like something small and ignorable. Hence, write it as:

$$e^2 \approx (3 - 0.28)^2 \approx 9 - 1.68 = 7.32$$

Which is within 0.1% of the actual value of 7.389... Another approach, depending on how you feel about squaring 27, would be to write:

$$e^2 \approx \frac{1}{100}(27 + 0.2)^2 \approx \frac{1}{100}\left[729 + 10.8\right] = 7.398$$

If it took you a while to calculate the square of 27, take a fresh look at the section "Square any two-digit number." (Add 3 to the 27 to make 30. Subtract 3 from 27 to get 24. Multiply 30 × 24 = 720. Then add on the square of 3, which is 9, to get 729.)

How it works

As $(a + b)^2 = a^2 + 2ab + b^2$, if b is small compared with a, ignoring the b^2 term increases calculation speed without damaging accuracy too much. If you think of 11^2, for example, we know that $11^2 = 121$. With this estimate, we'd write $11^2 = (10 + 1)^2$. and approximate this by 120, which is fairly close.

Try these

1. 54^2
2. 8.1^2
3. 0.19^2
4. 840^2
5. 8.7^2
6. 1.6^2
7. 51^2
8. 3.6^2
9. 570^2
10. 680^2
11. 41^2
12. 9.5^2
13. 0.44^2
14. 9.6^2
15. 0.13^2
16. 110^2
17. 8.7^2
18. 830^2
19. 0.49^2
20. 180^2
21. 51^2
22. 7.9^2
23. 71^2
24. 500^2
25. 8.1^2
26. 3.3^2
27. 42^2
28. 9.1^2
29. 490^2
30. 30.4^2
31. 3.03^2
32. 5.01^2
33. 80.1^2
34. 80.3^2
35. $6,940^2$
36. 0.693^2
37. 49.5^2

38. 498^2
39. 39.9^2
40. 7.95^2
41. 4.08^2
42. 740^2
43. 4.47^2
44. 5.42^2
45. 14.8^2
46. 2.61^2
47. 560^2
48. 1.68^2
49. 0.363^2
50. 6.62^2

Square roots, Babylonian style

Square roots are immensely useful things to calculate, especially in an engineering context. Many ancient civilizations had rough-and-ready ways to approximate square roots. These they used in engineering projects, such as building ramps, for example, to help carry things from the ground to the top of a steep wall. Rather than ponder construction projects, though, let's turn our thoughts to the body surface area (BSA) of a human being.

As every weight watcher knows, when you eat, you take in calories. When you exercise, you lose calories. You also lose energy, though, through the surface area of your body. If you are cooler than your surroundings, you give off heat by conduction, convection, radiation, and evaporation (sweating), and the amount of energy you transfer depends on your surface area. Those interested in human biology need a formula for BSA, and there are large numbers of competing equations from which to choose. BSA is not important merely for temperature regulation; there are also medicines for which the dosage is determined in part by the patient's surface area—such as several chemotherapy drugs. It's important, then, to have a formula that predicts BSA accurately. The simplest, mathematically, is Mosteller's formula, $BSA = \sqrt{MH}/6$, where M is mass in kilograms and H is height in meters. Special hospital beds can measure a patient's weight, and a patient, or their loved ones, can inform medical staff about their height, which is harder to assess if the patient is in great pain, tossing and turning to become comfortable.

To compute BSA, though not in a clinical setting, we look to one of the stars of the sumo wrestling world, Hakuhō Shō. Hailing from Ulaanbaatar, the capital city of the Mongolian People's Republic, he measures 1.92 m in height and tips the scales at an impressive 151 kg. Simplifying, as weight fluctuates in any given day, week, or month, let's call these 150 kg and 1.9 m, so that we underestimate our answer. A rapid multiplication gives MH in the Mosteller formula as 285, and this is what we need to take the square root of.

To find $\sqrt{285}$, we seek the highest perfect square closest to—but not exceeding—285. Those who know that is 16^2 have a head start. Our best initial guess, then, is that the square root is just over 16. We can do better. Construct an enhanced approximation by subtracting the 256 from the 285, dividing this by the square root of 256 (i.e., 16) and halving. That is to say:

$$\sqrt{285} \approx 16 + \frac{285 - 256}{2 \times 16} = 16\frac{29}{32} = \frac{541}{32}$$

This gives us $\sqrt{285} = 16\frac{29}{32} = 16.90625$, compared with the calculator value of $\sqrt{285} = 16.881943\ldots$, so the method works.

The improper fraction appears because, to finish off the calculation of Hakuhō Shō's BSA, we need to divide by 6. Again, as we don't need to be too exact, this becomes BSA $= 90/32 = 2.8125\ m^2$, though, given our approximations, 2.8 m^2 is close enough. This process can be iterated to give even more accurate approximations,[3] and is the precursor of the Newton–Raphson method. Indeed, the cuneiform tablet YBC7289 computes $\sqrt{2}$ correctly to about 6 decimal places.

One Ancient Babylonian baffler appears in the second millennium BCE cuneiform tablet VAT 6598, housed in the Berlin Museum. The Babylonians wrote on a clay surface and, when baked, it's rather like pottery. As a consequence, a good amount of Babylonian tablets have survived (they used a wedge-shaped stylus to mark marks into the soft clay, and as the Latin word for wedge is cuneus, such writing was called cuneiform). One problem posed on VAT 6598 (dated somewhere in the range 1800–1600BCE) is to find the diagonal of a door whose height is 40 and whose width is 10. Courtesy of Pythagoras's Theorem, we know this amounts to finding the square root of 1,700. A quick guess would be that

---

[3] You can combine simple electrical resistors in series and parallel, using the Babylonian method. See Carl E. Mungan and Trevor C. Lipscombe, "Babylonian Resistor Networks," *European Journal of Physics*, 33(3) (2012), 531–7.

$\sqrt{1,700} \sim 40$, but we can refine this slightly to see that $\sqrt{1,700} \geq 41$, as $41^2 = 1,681$. (see the section "Square any two-digit number ending in 1 or 9" for more details). Using the method, we refine our answer:

$$\sqrt{1,700} = 41 + \frac{1,700 - 1,681}{2 \times 41} = 41\frac{19}{82}$$

In decimal form, this is roughly 41.23. However, as we are engaging in mental mathematics, it's fine to leave it as a fraction. After all, the Babylonians, who counted in 60s, would have written their answer as a fraction, too.

How it works

$$(n + m)^2 = n^2 + 2nm + m^2$$

In our example, we know $(n + m)^2 = 67$ and we have set $n = 8$. If we assume m is small compared with n, so that we can ignore the term in m^2, we have:

$$m = \frac{(n + m)^2 - n}{2n}$$

As we know $(n + m)^2$. and n, we now have an approximate value for m, and thus for an improved estimate for the square root.

Square roots, Chinese style

Mathematicians in Ancient China had a relatively swift and easy way to calculate square roots of numbers less than 1 million. Here, it is easiest to use the theory (how it works) to show the method. We know that the square root of any six-digit number must have a hundreds, tens, and units column. Suppose we seek $\sqrt{589,486}$, and we call the answer $(100h + 10t + u)$. Squaring, we have:

$$589,486 = 10,000h^2 + 2,000ht + 100t^2 + 200hu + 20tu + u^2$$

First, inspect the number of tens of thousands on the left-hand side, which in this case is 58. The largest square closest to, but not exceeding, 58 is 49, whose square root is 7. Hence $h = 7$. Substituting in:

$$589,486 = 490,000 + 14,000t + 100t^2 + 1,400u + 20tu + u^2$$

Consequently:

$$99,486 = 14,000t + 100t^2 + 1,400u + 20tu + u^2$$

Now we seek the integer value of t such that $t \sim 99,486/14,000$, which gives $t = 7$. However, this is in fact too close, as the term $100t^2$ will also contribute to the 10,000 column and when combined with the $14,000t$ will be more than 99,486. Thus we hit on $t = 6$. Substituting this value in yields:

$$99,486 - 84,000 - 3,600 = 11,886 = 1,520u + u^2$$

Recall the way of regrouping numbers in order to add and subtract more swiftly. For example, we could rewrite this as $99,600 - 3,600 - 84,000 - 114$ which is fairly quick to compute.

To fix the value of u, we know that $1,520u$ is far bigger than u^2, so we evaluate $11,886/1,520 = 7$. Thus, our units column is 7. Or,

$$\sqrt{589,486} = 767$$

Which is extremely close to the actual answer of 767.7799, within 0.102%.

Square roots, Indian style

Mathematics has been recorded on papyrus in Egypt, inscribed in clay tablets in Mesopotamia, and written on tree bark in the Indus Valley. One Sanskrit mathematical manuscript dating to the third or fourth century CE was found in Bakhshālī, in 1881. The Bakhshālī manuscript, consisting of 70 leaves of birch bark, contains an intriguing way to calculate a square root. It outlines a method that we would write as follows. Suppose we seek the square root of a number N. Suppose further that m is the biggest number with $m^2 < N$. Form the remainder, r, which is $r = N - m^2$. The better approximation will be:

$$\sqrt{N} = m + \left(\frac{r}{2m}\right) - \frac{\left(\frac{r}{2m}\right)^2}{2\left[m + \left(\frac{r}{2m}\right)\right]}$$

Again, let us compute $\sqrt{67}$. We have $m = 8$, and so $r = 67 - 64 = 3$. Hence:

$$\sqrt{67} = 8 + \left(\frac{3}{16}\right) - \frac{\left(\frac{3}{16}\right)^2}{2\left[8 + \left(\frac{3}{16}\right)\right]}$$

Which we can recast as:

$$\sqrt{67} = 8 + \left(\frac{3}{16}\right) - \frac{3}{16}\frac{3}{32\left[8 + \left(\frac{3}{16}\right)\right]} = 8 + \left(\frac{3}{16}\right) - \frac{3}{16}\frac{3}{[256 + 6]}$$

So that:

$$\sqrt{67} = 8 + \left(\frac{3}{16}\right)\frac{259}{262} = 8\frac{777}{4,192}$$

To produce an answer in decimal form, I confess to using a calculator to find that it is 8.185353, impressively close to the calculator value of 8.1853527.

Kūshyār ibn Labbān, who was alive in 975 CE, wrote a mathematics text that was influential in the medieval Islamic world. In his work called *Principles of Hindu Reckoning*, the author asked the reader to come up with the square root of 65,342. Luckily, we have already learned how to square swiftly any number ending in a 5. Hence, we know that 255^2 is 65,025. This means $m = 255$ and so $r = 65,342 - 65,025 = 342 - 25 = 317$. Hence:

$$\sqrt{65,342} = 255 + \left(\frac{317}{510}\right) - \frac{\left(\frac{317}{510}\right)^2}{2\left[255 + \left(\frac{317}{510}\right)\right]}$$

Given the sizes of the terms involved, retaining only the first two terms should work well enough, and gives us $\sqrt{65,342} = 255\frac{317}{510}$. Alas, 317 is prime, so there's nothing much we can do to simplify our answer, which is approximately 255.62157 and therefore gobsmackingly close to the exact answer, 255.62081.

How it works

We begin with $N = m^2 + r$, and rewrite it as $N = m^2\left(1 + r/m^2\right)$. Take the square root of both sides to obtain $\sqrt{N} = m\left(1 + \frac{r}{m^2}\right)^{\frac{1}{2}}$. As we expect r to be smaller than m^2, we expand this keeping only the first few terms. That is to say, we apply:

$$(1 + x)^{\frac{1}{2}} = 1 + x - \frac{1}{8}x^2 + \frac{1}{16}x^3 \ldots \Big)$$

To obtain:

$$\sqrt{N} = m\left[1 + \frac{r}{2m^2} - \frac{1}{8}\left(\frac{r^2}{m^4}\right) + \left(\frac{r^3}{16m^6}\right) + \cdots\right]$$

$$= m + \frac{r}{2m} - \frac{1}{8}\frac{r^2}{m^3} + \frac{1}{16}\frac{r^3}{m^5} + \cdots$$

This expansion, if we continued it forever, would give us the exact result.

The Bakhshālī manuscript has:

$$\sqrt{N} = m + \left(\frac{r}{2m}\right) - \frac{\left(\frac{r}{2m}\right)^2}{2\left[m + \left(\frac{r}{2m}\right)\right]}$$

Rewrite as:

$$\sqrt{N} = m + \left(\frac{r}{2m}\right) - \frac{m}{2}\frac{\left(\frac{r}{2m^2}\right)^2}{\left[1 + \left(\frac{r}{2m^2}\right)\right]}$$

Again we expand, this time using:

$$\frac{1}{1+z} = 1 - z + z^2 - z^3 + \cdots$$

To get:

$$\sqrt{N} = m + \left(\frac{r}{2m}\right) - \frac{m}{2}\left(\frac{r}{2m^2}\right)^2\left[1 - \left(\frac{r}{2m^2}\right) + \left(\frac{r}{2m^2}\right)^2 + \cdots\right]$$

So that:

$$m + \frac{r}{2m} - \frac{1}{8}\frac{r^2}{m^3} + \frac{1}{16}\frac{r^3}{m^5} + \cdots$$

Which is what we had from our previous expansion.

Hence square roots, as per this 1,700-year-old method, coincide with the exact answer, provided the initial guess for the root is not too far off.

Square roots, French style

A French physician, Nicolas Chuquet, outlined in *Triparty en la science des nombres* an ingenious way to home in on square roots, something that in modern numerical analysis might be thought of as interval halving.

To approximate $\sqrt{67}$, one of our favorite numbers to take the root of, hem it in by two numbers. Clearly it must be more than 8, as $8^2 = 64$.

To save unnecessary effort, ponder deeply for a moment. We know $(8 + x)^2 = 64 + 16x + x^2$.. If $x = 1/8$, this would sum to less than 67. If $x = \frac{1}{4}$, it would sum to more than 67. We know, then, that:

$$8\frac{1}{8} < \sqrt{67} < 8\frac{1}{4}$$

To proceed further, there's an extremely useful formula to remember. Suppose we have a small fraction, a/b, and a larger fraction, c/d (both of which are greater than zero). Then:

$$\frac{a}{b} < \frac{a+c}{b+d} < \frac{c}{d}$$

This helps us choose the next approximation. That is to say, we know that $\frac{1}{8} < \frac{1}{4}$ and so we should next explore, using the formula:

$$\frac{1}{8} < \frac{1+1}{8+4} < \frac{1}{4}$$

The middle fraction is $1/6$. Consider, then, $(8 + 1/6)^2 = 64 + 16/6 + 1/36 = 66 + 4/6 + 1/36$. The nice thing is we don't have to calculate this, we just need to know whether it's large or smaller than 67. It's smaller. So we have:

$$8\frac{1}{6} < \sqrt{67} < 8\frac{1}{4}$$

The next step:

$$\frac{1}{6} < \frac{2}{10} < \frac{1}{4}$$

And $(8\ 1/5)^2 = 64 + 16/5 + 1/25 = 67\ 6/25$, leading to:

$$8\frac{1}{6} < \sqrt{67} < 8\frac{1}{5}$$

Our next guess is:

$$\frac{1}{6} < \frac{2}{11} < \frac{1}{5}$$

At this point, with $\left(8\frac{2}{11}\right)^2 = 64 + \frac{32}{11} + \frac{4}{121} = 60 + \frac{10}{11} + \frac{4}{21} = 66\frac{114}{121}$, we choose to stop. Our answer is $\sqrt{67} = 8.18181818...$, within 0.044% of the exact answer.

A couple of points. While the answer is pretty accurate, we can get even closer by continuing with the iteration. But it would have been far worse, say, had we have started with initial guesses of 8 and 9. Spending a few moments at the beginning to think of fairly close estimates is well worth the effort.

Albert Einstein was one of the greatest physicists ever, but his poorest subject in high school was French. Let's apply this French method, then, to one of Einstein's own calculations. In his doctoral thesis, Einstein came up with two expressions linking the number of molecules N in a mole of a substance, and the radius R of molecules of that substance (in meters). Putting in the numbers for a sugar solution, he got $NR^3 = 200$. and $NR = 2.08 \times 10^{16}$.[4] Dividing, we have $NR^3/NR = R^2 = 200/2.08 \times 10^{-16}$, so $R = 10^{-8}\sqrt{200/2.08} = 10^{-7}\sqrt{1/1.04}$. Let's apply Chuquet's approach. As we are taking the square root of a number less than 1, that number will be bigger than the square root itself. This gives us:

$$\frac{100}{104} \leq \sqrt{1/1.04} \leq 1$$

But we want a better upper bound. From the section "Multiply two numbers (or square a number) just over 100", we know that $1.01^2 = 1.0201$, which is less than 1.04. Therefore, we know that:

$$\frac{100}{104} \leq \sqrt{1/1.04} \leq 1\sqrt{1/(1.01)^2} \leq \frac{100}{101}$$

As the next step, form 200/205 as our next step, to obtain:

$$\frac{199}{204} \leq \sqrt{1/1.04} \leq \frac{99}{100}$$

And then on to:

$$\frac{298}{304} \leq \sqrt{1/1.04} \leq \frac{99}{100}$$

Aquiver with anticipation, we see that the left-hand side is $(300-2)/(300+4)$ and know that this is approximately $1-6/300 = 0.98$. Our estimate, à la Chuquet, is halfway between 0.98 and 0.99, which is 0.985. The radius of a molecule, based on Einstein's calculation, is

[4] See John Stachel (ed.), *Einstein's Miraculous Year: Five Papers that Changed the Face of Physics* (Princeton, NJ: Princeton University Press, 1998), p. 65.

9.85×10^{-8}. Einstein used 9.9 instead. He obtained $N = 2.1 \times 10^{23}$, but as his underlying calculation had mistakes in it, his answers for N and R were slightly off anyway!

Alas for Chuquet, his manuscript was not published in his lifetime and, to make matters worse, it was plagiarized by Estienne de la Roche, one of Chuquet's students. That said, de la Roche's book *l'Arismetique* was published in 1520, more than 20 years after Chuquet's death. At least the master did not live to see his student take his work and publish it without due credit.

Find the square root of a mystery perfect square

The Chinese square-root method, suitably adapted, permits swift calculation of the roots of perfect squares less than 10,000. Suppose you want to find the square root of 6,724. As with the previous section, ignore the last two digits, and look at 67. The closest perfect square to this is 64, which is 8^2. The tens digit is thus 8. The trickier part is to determine the last digit. As 6,724 ends in a 4, we look for a number that, when multiplied by itself, gives a 4. That could be a 2, but it could also be 8. The same dilemma holds for all other digits except for 0 and 5 (1 and 9; 2 and 8; 3 and 7).

To work out which to choose, think for a moment. We know $6,724 = (80 + n)^2$. with $n = 2$ or 8. We have, then, that $6,724 = 6,400+160n+n^2$. As n^2 is small compared with $160n$, we look at $324/160$, which is about 2, and so we plump for $6,724 = 82^2$.

Ever eager, let's try this again. What number, when squared, gives 3,364? We look at 33 and know immediately that the first digit of the number we want is 5. We then quickly do $3,364-2,500 = 864$. Divide $864/100$ and that gives us $3,364 = 58^2$.

Be warned, though. Taking square roots is not an area of arithmetic where we can play as fast and loose with decimal points and zeros as we have been. It's true that $\sqrt{6,724} = 82$ and that $\sqrt{67.24} = 8.2$, but $\sqrt{672.4}$ is something entirely different.

How it works

$$(10t + u)^2 = 100t^2 + 20tu + u^2$$

By looking at how many hundreds we have, we determine t^2 and thus t.

Find the cube root of a mystery perfect cube

This is more of a party trick than a practical piece of math, but it impresses an audience no end. The first step requires rote memorization. Take a look at the first nine perfect cubes. These are **1**, **8**, 27, **64**, 125, 216, 343, 512, and 729. The bold numbers provide a clue.[5] Namely, if you are given the cube 474,552, then as it ends in a 2 and the cube of 8 ends in a 2, we know this mystery cube must have 8 as its last digit. The next step is to remove the last three digits of the mystery cube and look at the remaining number, in this case 474. Looking at the list, you see the perfect cube closest to this number but below it (an important point!) is 343, which is 7^3. Our answer, then, is that $474,552 = 78^3$.

How it works

Setting the last three digits of the cube to zero generates a number that is $1,000n^3$. ($n^3 = 474$. in the example). The cube root of this number is $10n$. The cube root we seek, though, has the form $(10m + p)$. This means that m must be the integer closest to, but less than, n. This fixes m. But $\left(10m + p\right)^3 = 1,000m^3 + 300m^2p + 30mp^2 + p^3$ and as the numbers from 1 to 9 have distinct numerical endings when cubed, we can infer p.

Find the root of a mystery fifth power

This has a wow factor. Ask an audience member to think of a two-digit number and then (probably using a calculator) determine its fifth power. They tell you what the answer is, and you have to find out the fifth root. It is both true and useful that the last digits of the fifth power of the digits from 0 to 9 are the digits themselves. The next thing, should you wish to do this trick, is to learn by heart the fifth power of the following digits:

- $1^5 = 1$
- $2^5 = 35$

[5] To remember the numbers in bold, see that 1, 4, 5, 6, and 9 are "unchanged." The numbers behaving badly, at least in terms of cubing, are 2, 3, 7, and 8. But if you write these in the reverse order, you get 8, 7, 3, and 2. Putting this into the pattern gives 1, 8, 7, 4, 5, 6, 3, 2, and 9. Those who enjoy rowing have an added advantage, as numbers that remain unchanged are the important people in the boat: 1, 4, 5, 6, and 9 are the stroke, the engine room, and the cox, if it's a bow-loading shell.

- $3^5 = 243$
- $4^5 = 1024$
- $5^5 = 3,125$
- $6^5 = 7,776$
- $7^5 = 16,807$
- $8^5 = 32,768$
- $9^5 = 59,049$

The procedure follows that for the cubic powers. Ignore the last five digits of the number you've been given. Look at the remaining digits and write down the number that, when raised to the fifth power, is closest (but below) what you have. Inspect the last digit and insert, to get the answer.

For example, what number, when raised to the fifth power, is 992,436,543?

Ignore the last five digits, to obtain 9,924. The closest fifth power to this, without going over, is 7,776. Thus, it must be a number in the 60s. But, as the last digit of the mystery fifth power is 3, the last digit of this mystery number must be 3. So $992,436,543 = 63^5$.

Try these

1. Find the fifth root of 5,153,632
2. Find the fifth root of 2,219,006,624
3. Find the fifth root of 11,881,376
4. Find the fifth root of 28,629,151
5. Find the fifth root of 33,554,432
6. Find the fifth root of 345,025,251
7. Find the fifth root of 130,691,232
8. Find the fifth root of 4,084,101
9. Find the fifth root of 90,224,199
10. Find the fifth root of 6,590,815,232
11. Find the fifth root of 3,276,800,000
12. Find the fifth root of 3,939,040,643
13. Find the fifth root of 4,704,270,176
14. Find the fifth root of 371,293
15. Find the cube root of 148,877
16. Find the cube root of 778,688
17. Find the cube root of 941,192

18. Find the cube root of 32,768
19. Find the cube root of 10,648
20. Find the cube root of 140,608
21. Find the cube root of 300,763
22. Find the cube root of 357,911
23. Find the cube root of 13,824
24. Find the cube root of 117,649
25. Find the cube root of 970,299
26. Find the cube root of 24,389
27. Find the cube root of 830,584
28. Find the cube root of 110,592
29. Find the cube root of 9,261
30. Find the cube root of 250,047
31. Find the root of the perfect square 2,209
32. Find the root of the perfect square 8,649
33. Find the root of the perfect square 7,921
34. Find the root of the perfect square 4,225
35. Find the root of the perfect square 2,116
36. Find the root of the perfect square 3,844
37. Find the root of the perfect square 7,396
38. Find the root of the perfect square 2,601
39. Find the root of the perfect square 2,304
40. Find the approximate root of 80
41. Find the approximate root of 62
42. Find the approximate root of 72
43. Find the approximate root of 40
44. Find the approximate root of 29
45. Find the approximate root of 77
46. Find the approximate root of 42
47. Find the approximate root of 11
48. Find the approximate root of 37
49. Find the approximate root of 91
50. Find the approximate root of 31

7

Close-Enough Calculations: Quick and Accurate Approximations

He was a poet and hated the approximate.
RAINER MARIA RILKE, *The Journal of My Other Self* (Tr. John Linton)

Sir:
In your otherwise beautiful poem "The Vision of Sin" there is a
verse which reads—"Every moment dies a man, Every moment
one is born." It must be manifest that if this were true, the
population of the world would be at a standstill. In truth, the rate
of birth is slightly in excess of that of death.

I would suggest that in the next edition of your poem you
have it read—"Every moment dies a man, Every moment 1 1/16
is born."

The actual figure is so long I cannot get it onto a line, but I
believe the figure 1 1/16 will be sufficiently accurate for poetry.
CHARLES BABBAGE, tongue in cheek,
in a letter to Alfred, Lord Tennyson

There are times when a rough-and-ready answer is good enough. On
a test, for example, with time running out, quickly estimating an
answer—to help you eliminate a couple of potential answers—can be
of great benefit. In this chapter, we focus on a number of ways to come
up with approximate answers swiftly and, in some cases, show how a
little extra time can bump up accuracy greatly.

Divide by 9

The easiest way to come up with a best guess when dividing a number
by 9 is to multiply the number by 11. And multiplying by 11 is a swift
process.

For example, to calculate 234/9, first estimate this as 234/10, which means the answer should be more than 23.4. Form 234 × 11, which is 2,340 + 234 = 2,574. Putting in the decimal points, we have 234/9 ∼ 25.84. But we can do better. Shunt the numbers two places to the right and add again. We do this as division by 9 has the decimal value 0.11111 . . . By shifting our multiplication by 11 over two places and adding, we are on the road to getting 0.1111, which is much closer: 2,574 + 25.74 = 2,599.74. Given our estimate, we can write 234/9 = 25.9974 or, being bold, 234/9 = 26.

When discussing multiplication by 9, there was an application: converting a temperature from Celsius to Fahrenheit. The dog that didn't bark, to use Sherlock Holmes's phrase, was how to do the opposite. Now we can. To convert from degrees Fahrenheit (F) to Celsius (C) use:

$$C = (F - 32) \times \frac{5}{9}$$

The highest temperature recorded in the United States was at Greenland Ranch, California, on July 10, 1913, when the thermometer read 134°F. Greenland Ranch has since been renamed, aptly, as Furnace Creek and serves as the headquarters of Death Valley National Park. To convert this into a temperature that people fluent in the metric system can understand, subtract 32 from 134 to get 102. Multiply by 5 to get 510 and then divide by 9, which is the same as multiplying by 0.1111. The way to do that is to use rapid math to multiply 51 × 11, which is 561, then shunt over twice and add, to get 56.666°C.

Some Americans faced a bitterly cold day on January 23, 1971, when temperatures at Prospect Creek Camp, Alaska plummeted to a staggering −80°F. Following the rubrics, subtract 32 (take off 30, take off another 2) to get −112. Multiply by 5 to get −560 and then multiply by 0.11 to get −61.6. Shunt over twice and add to an extremely chilly −62.216°C. That's even without factoring in any wind chill factor.

While we still count in tens for most things, time and angles have a base-60 flavor to them. But what if time became metric? Enter KerMetric time, which divides 24 hours into 100 equal size units called, perhaps not surprisingly, Kermits. *The Muppet Movie*, starring Kermit the Frog, has a running time of 95 minutes. What's that, in terms of Kermits? A day consists of 24 × 60 = 1,440 minutes, which equals 100 Kermits. Thus, 14.4 minutes make a Kermit. *The Muppet Movie*, thus, runs for a total of 95/14.4 Kermits. Estimate as 100/20 = 5. Now consider 95/144 = 95/(9 × 16). This means we can now use our fast method for division by 9. We know that 95/9 can be written approximately (ignoring decimals!)

as 95 × 11, which from "Multiplication by 11" is 1,054. Shunting this answer three places to the right gives an extra 1.054, and adding this gives the better answer (95/9 = 95 × 1.111) of 10.54 + 1.054 = 10.5554. Repeating this procedure suggests an answer of 10.555555. Now divide by 16, which involves halving, a lot!

This gives 95 minutes as 6.597 Kermits.

How it works

The decimal expansion of 1/9 is 0.1111. The method works because we truncate this to 0.11 and, as we always worry about the decimal points later, division by 9 has now become multiplication by 11. We know that the exact answer will be more than the one given by this method, as we're truncating 0.1111 after the first two decimal places. We also know that 2/9 = 0.2222, 3/9 = 0.333... and so forth. Thus when we obtain 25.84, we know the actual answer must be more than this and, consequently, must be 25 8/9.

Divide by 11

Division by 11 is a close relative of division by 9. If we wish to find 234/11, we estimate this as 222/11 so the answer exceeds 20. Multiply the number by 9: 234 × 9 = 2,340−234 = 2,106. We now write:

$$2 \quad 1 \quad 0 \quad 6$$
$$2 \quad 1 \quad 0 \quad 6$$
$$2 \quad 1 \quad 0 \quad 6$$

As many times as one wishes. That is to say, we take nine times the number and repeat it, shunting to the right by two places. Adding:

$$2 \quad 1 \quad 2 \quad 7 \quad 2 \quad 7 \ldots$$

Inserting the decimal point, we obtain the approximate answer 234/11 = 21.2727... We know the exact answer is 21 3/11.

Leonardo of Pisa, also known as Fibonacci, wrote *Liber Abaci* in 1202. This book introduced into the Latin West a number of mathematical techniques developed by Islamic mathematicians, such as al-Khwārizmī (from whose name the computer science term "algorithm" is derived). It also made great use of Indo-Arabic, rather than Roman numerals. One of the questions the book asks and answers is how to divide 12,532 by 11. Following our quicker calculation method, we first estimate the answer, which is about $12{,}100/11 = 1{,}100$. Thus we form $9 \times 12{,}532 = 112{,}788$ and move this over by *two* spaces several times and sum:

1	1	2	7	8	8				
		1	1	2	7	8	8		
				1	1	2	7	8	8
1	1	3	9	2	7	1	5	8	8

Putting in the decimal point, we have $12{,}532/11 = 1{,}138{,}271{,}588$ or, as we can infer the fraction, $1{,}139 \, 3/11$.

As with 7, there is an intriguing way to determine if a large number is exactly divisible by 11. To do so, you add up the digits of the number, but with alternating signs. For 82,654, for example, we sum $8-2+6-5+4=11$ and, as this is divisible by 11, so, too, is 82,654.

How it works

The decimal expansion for $1/11$ is $0.09090\ldots$ Truncating this gives 9, neglecting as usual the decimal points. Thus multiply by 9, shift the answer over two places, and over another two places. . . and then add. As the expansion for $1/11$ is so easy, one knows immediately what $2/11$, $3/11$, and so on look like, allowing us to go from an approximation to the actual answer, in fraction form.

Divide by 13

To find a rough expression for division by 13 is a slightly more complicated process than for 9 or 11. But not much more. It's based on the fact that $77 \times 13 = 1{,}001$. All you need to do to divide by 13, then, is to multiply by 77 (or rather, by 7 and then use the trick for rapidly multiplying by 11) and add in the right number of zeros.

To show the method, consider $234/13$ and estimate it as $260/13$, so we expect an answer of less than 20, but closer to 20 than to 10.

Multiply $234 \times 7 = 1{,}638$. And now multiply by 11, which is $10+1$, to get:

$$\begin{array}{r} 16{,}380+ \\ \underline{1{,}638} \\ 18{,}018 \end{array}$$

Inserting the decimal points, we get $234/13 = 18.018$. The exact answer is $234/13 = 18$.

Another problem posed in Leonardo of Pisa's *Liber Abaci* is to calculate 123,586 divided by 13. This should be less than $130,000/13 = 10,000$. Multiplying by 7 gives $123,586 \times 7 = 865,102$. Multiplying by 11 results in $8,651,020 + 865,102 = 9,516,122$. Inserting the decimal points, we get $123,586/13 = 9,516$, about 0.1% above the calculator value of 9506.6153.

To see if a number is an exact multiple of 13, remove its last digit and multiply by 9. In the case of 234, we take off the 4, multiply by 9, to obtain 36. We subtract this from what remains of the original number. Thus we form $23-36 = -13$. As this is exactly divisible by 13, so is 234.

How it works

We write $\frac{n}{13} = \frac{77n}{13 \times 77} = \frac{77n}{1,001} \approx \frac{77n}{1,000}$. The approximation will always be slightly higher than the true value.

Divide by 14

In the two previous sections, we saw how to divide, approximately, by 9 and 11. Notice that $9 \times 11 = 99$, which is close to 100. Another pair that resembles this is $7 \times 14 = 98$. To divide by 14 approximately, you need only multiply by 7 and insert the decimal points.

For example, $234/14$ is less than $280/14 = 20$. So we expect the answer to be less than 20. However, $234 \times 7 = 1,638$. The approximate answer, then, is 16.38. The exact answer is 16.714... As the approximation is fueled by the fact that $7 \times 14 = 98$ is roughly 100, we'd expect to be about 2% too low on our answer. If you double the 16.38 to get 32.76 and shunt over two places to obtain 0.3276 and add it to the first estimate, you get $234/14 = 16.38 + 0.3276 = 16.7076$, which is closer.

Cucumber lovers may know that one cup of sliced cucumber possesses 14 calories. If you have a 2,000 calorie a day diet, this means you can eat 2,000/14 cups of cucumber. This should be less than 200 but more than 150. Multiplying by 7 gives us the answer of "about" 140 cups. The measuring term "cup" in America has, as its metric equivalent, 250 milliliters, which is $\frac{1}{4}$ of a liter. Hence you can eat 35 l of cucumber per day, if you so choose. Bon appétit!

How it works

We seek $A = n/14$:

$$A = \frac{7n}{14 \times 7} = \frac{7n}{98} = \frac{7n}{100 - 2}$$

Now make use of the formula:

$$\frac{1}{1 - x} = 1 + x + x^2 + \ldots$$

To see that:

$$A = \frac{7n}{100}\left[1 + \frac{2}{100} + \ldots\right]$$

By multiplying by 7, we generate $7n/100$. By doubling, and dividing by 100, we add on the second term, which gets fairly close to the exact answer.

Multiply or divide by 17

To multiply by 17, approximately, observe shrewdly that $17 \times 6 = 102$. As a consequence, all you need to do is divide by 6 and add the right number of zeros at the tail.

Those jolly pranksters at the Massachusetts Institute of Technology, mercifully abbreviated to MIT, decided in 1958 that if Oliver Reed Smoot, Jr., wanted to join the Lamba Chi Alpha fraternity, he would have to lie down repeatedly on the Harvard Bridge and measure the bridge in terms of his own body length.[1] The answer was 364.4 Smoot body lengths. Alas, this is not a standard metric measurement. But Smoot was about 1.7 m tall. Hence, we know the bridge is roughly 364.4×1.7 m in length, so about 600 m long. We now ignore, as per our usual pattern, the decimal points. Take the 3,644 and halve it, to get 1,822 and then divide by 3, to get 607.33 m. As this is the right order of magnitude, it must be our answer. The actual span is 620.1 m. In honor of this feat, a commemorative plaque was placed on the bridge in 2008 and, in 2011, the "smoot" became a word included in the *American Heritage Dictionary*. His cousin bested him, for in 2006 George Smoot won

[1] Those who play Grandmother's Footsteps (or Mother, May I?) will recognize that Smoot moved across the bridge one "lamp post" at a time.

the Nobel Prize in physics for his work on the cosmological microwave background radiation.

Of a slightly more practical nature (though not by much!), pioneering French science fiction author Jules Verne wrote *20,000 Leagues Under the Sea*. A league, though, is meant to be the distance you can walk in an hour, and so is normally thought of as a unit equal to 3 miles. At sea, it transforms into 3 nautical miles, which is 3.452 miles. By now, we spot this as roughly 17×203, with some decimal points thrown in. How far below the surface is a mere 13,000 leagues? Clearly, it's more than 39,000 miles. To be more exact, strip off the 13, and multiply it by 203, which you can write down immediately as 2,639. Now, to multiply by 17, we divide by 6 to get 43,983. Our prior estimate implies that 13,000 leagues is 43,983 miles, which is within 2% of the correct answer (about 44,880 miles or 111,120 km). Scientists assure us that Challenger Deep, the deepest part of the ocean, is just under 11 km below the surface.

The highest (legal) speed limit in the United States is a 41-mile (66 km) stretch of Texas State Highway 140, dubbed the Pickle Parkway. This connects San Antonio to Austin and the top speed is 85 mph (137 km/h). To travel this section of road takes a time 41/85 hours, which is 82/170 hours. As multiplication by 17 requires us to divide by 6, division by 17 requires us to multiply by 6. We get $6 \times 82 = 492$ and as our answer must be just under half an hour, it'll take 0.492 hours to drive this part of the Pickle Highway. Multiply by 6 once again to obtain 29.52 minutes.

There's also a division rule for 17. To see if a number is a multiple of 17, remove the last digit of the number and multiply by 5. For 234, then, we take off the 4, multiply by 5, to obtain 20. We take this number from what remains of the original number, 23, to get $23 - 20 = 3$. As this is not divisible by 17, neither is 234.

Divide by 19 or 21

The easiest way to estimate either of these cases is to say that 19 and 21, roughly, are 20. Suppose we want 3,456/19. Estimate as $3,800/19 = 200$. Pretending that it is 20 gives $3,456/20 = 172.8$.

We can do better. Divide this answer by 20 again, to get 8.64. Add the 172.8 and the 8.64 to get $172.8 + 8.64 = 181.44$. Our approximate answer for 3,456/19 is therefore 181.34, not too far from the actual answer of 181.894737... Our answer is lower, and always will be. But this is an iterative (or rinse and repeat) method. We could take that 8.64 and divide that by 20 to get 0.432 and add that on to get 181.872, accurate to within 0.015%.

Why might you want to divide by 19? Albert Einstein might show the way. He wrote a note to a close friend of his first wife, addressing it to "Dear Mrs. Savić ($+1/9 \frac{1}{2}$ hopefully)!"[2] To be 4 months pregnant, to use Einstein's estimate, is to be 8/19 toward having a baby. We'd estimate this as about $0.4 + 0.02 + 0.001$, which gives 0.421, not too far from 0.4210526... from the calculator.

To divide by 21, you need to subtract, rather than add. Hence, we get $3,456/21 = 172.8 - 8.64 = 164.16$, close to the actual value $3,456/21 = 164.571429...$ When you iterate, you need to flip the sign. The next step would be to add 0.432 back in, to get 164.592, also accurate to within 0.15%.

To see if a number is an exact multiple of 19, double the last digit and add it to what remains of the original number. For 8,265, for example, remove the 5, double it to form 10, and add it to 826 to get 836. This is divisible by 19, and thus so is 8,265. If it's not clear that 836 is indeed a multiple of 19, repeat the process: $83 + 12 = 95$, which is 5×19.

In a similar way, to see if a number is exactly divisible by 21, double the last digit but now subtract it from what remains. For 6,174, remove the 4, double it to obtain 8, and subtract from 617, which yields 609. Repeat the process, which generates $60 - 18 = 42 = 3 \times 21$. As this is divisible by 21, so is 6,164.

How it works

Use the expansion $\frac{1}{1-x} = 1 + x + x^2 + x^3 + \ldots$, for when x is small, keeping only the first two terms. That is to say, we write $\frac{3,456}{19} = \frac{3,456}{20(1-\frac{1}{20})} = \frac{3,456}{20} + \frac{3,456}{20.20} + \ldots$ or alternatively $\frac{1}{1+x} = 1-x+x^2-x^3+\ldots$, so that $\frac{3,456}{21} = \frac{3,456}{20(1+\frac{1}{20})} = \frac{3,456}{20} - \frac{3,456}{20.20} + \ldots$

Divide by 24 or 26

Similar to the previous sections, we can make a first stab that division by 24 or 26 is going to be close to 25 (and, for a quicker calculation, requires multiplication by 4).

---

[2] Milan Popović (ed.), *In Albert's Shadow The Life and Letters of Mileva Marić, Einstein's First Wife* (Baltimore, MD: The Johns Hopkins University Press, 2003), p. 71.

Estimate 3,456/24 as roughly 3, 500/25 = 140. For greater accuracy:

$$\frac{3{,}456}{26} = \frac{3{,}456}{25+1} = \frac{3{,}456}{25}\left(1 - \frac{1}{25}\right) = \frac{3{,}456}{25} - \frac{3{,}456}{25.25} + \ldots$$

To compute 3,456/25, double the 3,456 twice and divide by 100. Doubling 3,456 gives 6,912, which in turn doubles to 13,824, so 3, 456/25 = 138.24. To include the second term, to add to precision, divide this once again by 25, and so we double 13,824 to get 27,648 and repeat the process to get 55,296. We know this must have the decimal point inserted to give 5.5296. This process gives our quick answer:

$$\frac{3,456}{26} = 138.24 - 5.53 = 132.71$$

Which is close to the answer of 132.92 (some of the decimal places were lopped off, simply because this is an estimate).

If you are asked what 3,456/24 happens to be, you could go through this process, which will give you an answer of 143.77. If you stare at the equation long enough, though, you'll realize that it's an exact integer: 3,456/24 = 144. Sometimes it pays to look closely first, rather than jump into using a method.

Try these

1. 997/9
2. 3.49/1.1
3. 99.8/13
4. 9.8/140
5. 70.8/1, 700
6. 56.5/19
7. 4.85/0.21
8. 83.7/2, 400
9. 800/26
10. 57.6/9
11. 63.4/1.4
12. 82.3 × 0.17
13. 56.4/130
14. 6.54/2, 600
15. 71.2/1.1
16. 0.263/1.3

17. 94.3/0.24
18. 71.8/1, 900
19. 14/17
20. 81.3/0.013
21. 7.22/1.9
22. 316/0.021
23. 280 × 170
24. 79.5/0.19
25. 3.22/14
26. 569/210
27. 47.6/0.9
28. 621/110
29. 1, 710 × 0.017
30. 58.8/2.6
31. 837/90
32. 750/0.11
33. 3.72/0.017
34. 6.36/190
35. 5.08 × 1.7
36. 35.6/260
37. 60.6/170
38. 5.64/1.1
39. 21.5/140
40. 97.9/2.4
41. 1.78/1.3
42. 9.37/21
43. 845/9
44. 5, 380/1.4
45. 6.53/1.7
46. 1.5/240
47. 65.1/2.1
48. 1.2 × 17
49. 6.22/24
50. 24.7/0.26

Multiply or divide by 32, 33, or 34

As with the complementary pairings of 9 and 11, 7 and 14, and 13 and 7, the numbers 32, 33, and 34 play nicely with the number 3. To multiply by 32, 33, or 34, simply divide the number by 3 and insert a decimal point in the right place. To divide by 32, 33, or 34, multiply by 3 and add zeros if necessary.

Suppose we want 476 × 33, for example. We simply divide 476 by 3 to get 158.6666 and as the answer must be about 12,000, we suspect 15,867. Or, rather, that's our first stab at the answer. What we do is to shunt this two places to the right and subtract. As in, $15, 867 - 158 = 15, 709$, which turns out to be the exact answer (the method isn't exact; we stumbled on to the right answer by ignoring the extra decimals, recognizing we could end up just a bit off).

If you are interested in multiplying by 32 or 34, you could do something flashy with series expansions. Or you could just take your answer for multiplying by 33, and add or subtract the number as appropriate.

The longest game in professional baseball (so far) pitted the Pawtucket Red Sox against the Rochester Red Wings, Triple-A affiliates of the Boston Red Sox and the Minnesota Twins, respectively. The game took place over 3 days (no big deal for cricket fans), during which a monster 882 pitches were thrown before the Red Sox triumphed 3–2. How many pitches per inning is that?

To find 882/33, we multiply 882 × 3 to get 2,646. Divide this by 100 to get 26.46 and add, obtaining 2,672.46. The answer has to be about 990/33 = 30, so we propose 882/33 = 26.7284 pitches per inning. As a check, the calculator displays 26,72727. . . and we are accurate to within 0.01%.

To divide by 32 or 34 instead, take the 26.7284 and multiply by 3 and divide by 100, which gives 0.801852. Add to the 26.7284 to get 882/32 = 27.530252. To divide by 34, subtract the 0.801852 instead, giving 882/34 = 25.926548. The calculator values are 27.5625 and 25.9411765. . ., respectively, so our answers are good.

How it works

Write $33n = \frac{99n}{3} = \frac{n}{3}(100 - 1) = \frac{100n}{3}\left(1 - \frac{1}{100}\right)$. Hence, we take the number, divide by 3, and add two zeros to get our first answer, then shunt over two places to the right and subtract to get an improved answer.

To divide we have:

$$\frac{n}{33} = \frac{3n}{99} = \frac{3n}{100 - 1} \approx \frac{3n}{100}\left(1 + \frac{1}{100}\right)$$

To divide by 32, we write instead:

$$\frac{n}{32} = \frac{n}{33-1} \approx \frac{n}{33}\left(1 + \frac{1}{33}\right) \approx \frac{3n}{100}\left(1 + \frac{1}{100}\right)\left(1 + \frac{1}{33}\right)$$

This, though, is $\left(1 + \frac{1}{33}\right) \approx 1.03$ multiplied by our answer for division by 33. Hence, we take our previous answer and increase it by 3%.

Multiply or divide by 49 or 51

The easy approach here is, as with our case of 19 or 21, simply to assume that multiplying or dividing by 50 is good enough. The answer will be better—closer to the actual answer—since the percentage errors will be smaller.

For example, 3,456/49 or 3,456/51 is approximately 69.120, as opposed to the actual values of 70.53. . . and 67.76. . ., respectively.

However, we know that:

$$\frac{3,456}{49} = \frac{3,456}{50\left(1 - \frac{1}{50}\right)} = \frac{3,456}{50} + \frac{3,456}{50.50} + \cdots$$

Hence, we can take the 69.12 and divide by 50 once more, to obtain 1.3824 and then add to get:

$$3,456/49 = 69.12 + 1.38 = 70.5\ldots$$

And:

$$3,456/51 = 69.12 - 1.38 = 67.74$$

Which are extremely close to the actual values.

The Azerbaijan Grand Prix is a race for Formula One cars that consists of 51 laps. The length of the race is 306.049 km, so what is the length of one lap? We seek 306.049/51 = 6.12098–0.1224196 = 5.9985604 km. To this, we tack on the next term, which contributes 0.002448294 km, to give 6.001 km. As the track is 6.003 km, our answer is off by about 2 m, less than half the length of a typical Formula One racing car.

Multiply or divide by 66 or 67, 666 or 667, Newton's Universal Constant of Gravitation or Planck's constant

Here is wisdom. Let him that hath understanding count the number of the beast: for it is the number of a man; and his number is Six hundred threescore and six.

Revelation 13:18 KJV showing the usefulness of the number 666

To divide by 66 or 67, simply multiply the number by 3, divide by 2, and insert the appropriate decimal points.

Consider 3,456/66, which we estimate as $3,000/60 = 50$.

Halve 3,456 to get 1,728. Then multiply by 3 to get 5,184, and insert a decimal point to get $3,456/66 = 51.84$.

To multiply by 66 or 67, do the opposite. Multiply by 2 and divide by 3. This is straightforward, but may be of some use to physicists. Newton's gravitational constant $G = 6.674 \times 10^{-11} \text{N kg}^{-2} \text{ m}^2$, and so this method helps to find out multiples of G swiftly. Switching from cosmos to quantum, Planck's constant $h = 6.626 \times 10^{-34}$ Js, and so the same method can help there as well.

Horse racing, so the saying goes, is the sport of kings. Dog racing is a sport of the people. On the evening of January 20, 2011, fans of dog racing were awed by an amazing feat. Three dogs tied in a triple dead heat for first place in the longest event, the 925 m at Romford Stadium. Greyhounds have a top speed of 67 km/h, so we can estimate how long it took. The time is the distance, 925 meters, divided by speed. To convert kilometers per hour to meters per second, we divide by 36 (as 1 km/h is 1,000 m in 3,600 seconds). Our estimate for the time is therefore $925 \times 36/67$ and we know that $925 \times 36 \times 3/2$ is close enough. Our answer is 925×54. Multiplication by 54 is something we can also do swiftly, as $54 = 50 + 4$, and so we need only multiply by 50 (as in, divide by two) to get 46,250 and then double 925 to get 1,850 and double again to get 3,700. Our answer is now $46,250 + 3,700 = 49,950$. As the dogs were racing about 1 km and run at approximately 60 km/h, the race should have lasted less than a minute. The race, we predict, should take about 49.95 seconds. In fact, the dogs stopped the clock at 59.53 seconds.

Our estimate was good, given that we used the top speed of a greyhound, the so-called "45 mph couch potato" of the canine world. A rematch took place a week later and Droopy's Djokovic—the rank outsider a week earlier—triumphed.

At the time, no-one could recall a triple dead heat in dog racing occurring before. But in one of those "truth is stranger than fiction" stories, fans didn't wait long for a second. On December 13, 2012 (now that you've read Doomsday, what day of the week was that?), Wimbledon dog track saw a three-way tie for first place in a 480-m race. Following the same procedure as before, we expect a time of about 25.92 s. The winning time was 28.87 seconds. Perhaps obviously, our estimate is more accurate for the shorter race than the longer one!

Divide by 91

While we have relied upon pairs of numbers that, when multiplied, are close to 100, here we go to 1,001, given that $91 \times 11 = 1,001$. So, to divide by 91, you can get very close by multiplying by 11, and we already know a method for swiftly multiplying by 11. In passing, the fact that 1,001 is the product of three primes, $1,001 = 7 \times 11 \times 13$, makes it a so-called sphenic number. Construct some sphenic numbers on your own (to get you started, the smallest is $2 \times 3 \times 5 = 30$; there are only five of them less than 100).

For 876/91, then, we estimate the answer as 10. Then multiply 876×11, which we write down immediately as 9,636 and so know that we must have 9.636 as our answer. We can do better. We know that $\frac{1}{1,001} = \frac{1}{1,000}[1 - \frac{1}{1,000} + \dots]$ and so a better estimate will be $9.636 - 0.00936 = 9.62664$, close to the actual answer of 9.6263736...

Not just a sphenic number, 91 is also a cannonball number or, more formally, a square pyramidal number. If you see cannonballs stacked, there's one on top, four underneath, nine underneath that, and so on. How many balls do you need to make a stack? One would work. So too would $1 + 4 = 5$. Or $1 + 4 + 9 = 14$, or $1 + 4 + 9 + 16 = 30$. The numbers 5, 14, and 30 are square pyramidal numbers. The next two are 55 and 91. With 91 cannonballs, or oranges, or apples, you could make a pyramid of six layers.

What if you had 4,096 oranges you wanted to arrange in six-layered pyramids, how many stacks would you have? We require 4,096/91 and so, following the recipe, we perform $4,096 \times 11$, which gives 45,056. We expect an answer of about 4, 000/100 = 40, so we might write our first attempt as 45.056. Our enhanced answer, once we subtract 0.045056, is 45.010944. There is some more information here, though. As the series has alternating signs, and each successive term is far smaller than the one that preceded it, we know not only that 45.056 is too high, but that 45.01044 is too low. How low? We know 45 × 91 is (90 × 91)/2 = 8, 190/2 = 4, 095. There are 4,096 oranges, which will form 45 stacks, leaving one left over. Any sensible person would eat the last orange.

Sir Walter Raleigh posed the problem of how to stack cannonballs most efficiently to Thomas Harriot. He solved the problem, but mentioned it to Johannes Kepler, who in 1611 turned it into the Kepler Conjecture,[3] that the way in which you stack cannonballs, or fresh fruit, is the densest way to pack spheres that exists—the largest number of spheres in the smallest possible volume. Or, to be more precise, there is no denser packing possible. A formal proof was published in 2017, some 406 years after Kepler proposed his conjecture.

Divide by 111

As in the previous section, we now look at numbers close to 1,000. As 9 × 111 = 999, we know that dividing by 111 is approximately the same process as multiplying by 9. To find 926/111 note the answer is about 9. Write down:

$$9, 260–926 = 8, 334$$

Our prior estimate suggests the answer 8.334. But as 999 = 1, 000–1, we can improve our answer by adding to 8.334 the number 0.008334, to get 8.342334, close to the calculator value of 8.3423423. . .

Divide by 198, 202, and other multiples of 99 and 101

Again we open up our can of series expansions. If we take 495, for example, and want to find 113/495, we could write:

$$\frac{113}{495} = \frac{113}{500 - 5} = \frac{113}{500\left(1 - \frac{1}{100}\right)} = \frac{113}{500}\left(1 + \frac{1}{100} + \dots\right)$$

[3] It appears in Kepler's book *The Six-Cornered Snowflake* (in Latin, *Strena seu de Nive Sexangula*) Strena is New Year, and the book was intended as a New Year's gift for a friend, hence the more-formal title *New Year, or the Six-Cornered Snowflake*.

We think of 1,130/5, which gives us 226, shunt over two places to get
228.26 and, as we know the answer must be about 0.2, we have 113/495 =
0.22826. To add another term and become more accurate, shunt over
by four places, to get 113/495 = 0.2282826. The calculator yields
0.22828283.

Speaking of 495, consider the following. Take any three digits (not all
the same), say 314. Now use them to write the down the biggest and
the smallest number you can, in this case, 431 and 134. Subtract the
smaller from the larger, which is 297. Do it again! 972–279 = 693. And
again, so that 963–369 = 594. And last, 954–459 = 495. If you want
to keep going, and why not, you have 965–459 = 495, so this is where
we stop. You can now try this with four digits, and you'll end up with
6,174 instead of 495. The number 6,174 is called the Kaprekar number,
and Kaprekar numbers exist for different bases, as well as for different
numbers of digits.[4]

Returning to division and using the same reasoning (series expansion) as before, we know:

$$\frac{113}{505} = \frac{113}{500 + 5} = \frac{113}{500\left(1 + \frac{1}{100}\right)} = \frac{113}{500}\left(1 - \frac{1}{100} + \dots\right)$$

So that 113/500 = 0.22374. Adding that extra term yields 0.2237626, with
the calculator notching up 0.22376238. Not bad!

Try these

1. 2.11/0.33
2. 30.1 × 490
3. 68.1/5.1
4. 7.72/660
5. 40.3 × 3,300
6. 990 × 4.9
7. 52.1/6.7
8. 7.18 × 0.0033
9. 9.31 × 510
10. 9.23/66.6
11. 3.24/1.11

[4] D.R. Kaprekar, "On Kaprekar numbers," *Journal of Recreational Mathematics*, 13
(1980–81), 81–82.

12. 8.04×666
13. $19.3/3.3$
14. 71.1×5.1
15. $0.676/0.67$
16. 4.12×0.34
17. $57.9/49$
18. $4.37/6.6$
19. $240/0.111$
20. $52.2/33$
21. 0.0066×86.8
22. $40.2/340$
23. 670×0.855
24. 56.7×0.033
25. $47.9/34$
26. $830 \times 6,670$
27. $0.740/66$
28. 51.3×3.4
29. $9.51/0.51$
30. 7.6×0.049
31. $7,430/666$
32. 4.02×6.7
33. 3.34×340
34. $4.47/0.91$
35. 66×5.26
36. $9.62/0.49$
37. $1.39 \times 6,660$
38. $86/11.1$
39. 67×8.46
40. $110/4.9$
41. $461/51$
42. 36.4×0.066
43. $9.51/67$
44. $3.52/9.1$
45. 667×481
46. $260/6.66$
47. $470/91$
48. $38.1/0.667$
49. $6.91/66.7$
50. $3.14/0.34$

Interlude V

Approximating the Number of Space Aliens

> I'm sure the universe is full of intelligent life. It's just been too
> intelligent to come here.
>
> ARTHUR C. CLARKE (reproduced from an interview
> http://www.scifi.com/transcripts/aclarke.txt)

The Vietnam War, during which American casualties ran extremely
high, remains controversial in the United States. During the conflict,
US forces estimated the strength of enemy forces based on the "SWAG"
principle. At the war's end, in a legal case, Colonel John Stewart took the
stand. Lawyers grilled him, asking what, exactly, SWAG stood for. His
reply, generating much amusement in the courtroom, was "Scientific
Wild-Ass Guess."

A SWAG, though, is not an opinion. Or, rather, it is—but not as we
often think about things. You and I might disagree about who is the
greatest soccer player of all time (the correct answer, surely, is Derek
Hales, once of Charlton Athletic). But it's an area in which your opinion
and mine are probably equally valid. Should an expert offer an opinion
within their area of expertise, their views should have greater weight
than non-experts.[1] In other words, anyone can make a wild-ass guess,
but a *scientific* wild-ass guess is of a higher caliber. Arguably the greatest
SWAG came from Saint John XXIII. According to a legend—which
given the Pope's well-known sense of humor could well be true—a
news reporter asked him "How many people work at the Vatican?" to
which he replied "About half of them."

---

[1] Social media place everyone's opinion on an equal footing, so it seems. As a cogent,
persuasive warning, see Tom Nichols, *The Death of Expertise: The Campaign against Established
Knowledge and Why it Matters* (New York: Oxford University Press, 2017).

The scientific wild-ass guess differs from a mere wild-ass guess. Scientific implies some thought goes into the estimate, or it's made by someone with specialist knowledge who might reasonably know what they are talking about. A good example is a protest march, complete with a counter-demonstration. The protest organizers, when speaking to the media and on their own social media sites, will probably greatly overestimate the number of marchers. The counter-demonstrators will significantly underestimate the number of marchers and, no surprise, overestimate the number of counter-demonstrators. Oddly enough, both sides will produce photographic "evidence" to support their numbers. A photo taken at the head of the march, where folks crowd in to be next to the Very Important People marching tells one story; a video taken by counter-demonstrators of the end of the march as stragglers pass them by conveys an entirely different one. The police, whose job it is to protect both groups, and to take care of marchers and counter-demonstrators every weekend of the marching season, are the ones who provide a SWAG, rather than numbers based on wishful thinking. In Washington DC, there used to be two SWAGs provided for those who marched on the National Mall: one from the police, the other from the National Park Service. The two numbers often differed considerably: in the end, a SWAG, no matter how scientific, is still a guess.

Leveling up from a SWAG brings us to order-of-magnitude estimates. To do so, toss aside the uninteresting numbers to leave only the important ones. The idea here is to generate an answer that is "about a million" "about a hundred thousand," "about ten." If you want to discover a new elementary particle and shatter the standard model of particle physics, you'll need to come up with an order-of-magnitude estimate for its mass and energy so that the world's atom smashers know where to look for it.

Let's use order-of-magnitude estimates to put on a show! Suppose I claim to have psychic powers (I don't; they don't exist) and a TV show that usually has 600,000 viewers invites me to demonstrate my uncanny unearthly abilities. Placing my hands on my forehead, I concentrate hard and tell viewers that I am focusing my mental energies on the light bulbs in their house. If I manage to cause one of their light bulbs to go out (I'd better be appearing on an evening show!), I want them to call, text, or tweet the TV station. Suddenly, the switchboard and the internet are full of messages. My amazing powers have been confirmed! But not really. An incandescent light bulb lasts between 1,000 and

2,000 hours. To make the math easier, call it 1,500 hours. With 600,000 viewers, during an hour-long show there should be 600,000/1,500 bulb deaths, or about 400. That should generate more than six messages every minute during showtime. And if everyone watching has two lights on, I've just doubled the number of outages. A simple order-of-magnitude estimate shows how such performers succeed.[2]

Now I turn my non-existent psychic powers to those who have passed on to the other side. If I focus on the dead while in front of a live audience, I could claim to have a weird vision of a funeral—one that involved a rhino. Again, I know that a studio audience has, say, 300 people in it. Each member probably knows 200 people each, which puts me in touch with 60,000 people. I sense someone from the other plane contacting me, the vibrations aren't clear, someone who passed away and—this seems strange—something to do with a rhino. Does anyone here connect losing a loved one and a rhino? This is known in the trade as a Barnum statement. It's absolutely crazy to associate a funeral with a rhino, but I'm speaking to 300 people and, via them, to 60,000 people. Perhaps an audience member's relative was killed by a rhino—a big plus for my TV career. Or someone was buried wearing their favorite t-shirt with a rhino on it. Or they were buried in a wildlife-themed coffin (a cardboard coffin with an elephant safari scene on it retails for just under £400). An audience member tells me their uncle recently died and was a giraffe keeper at a zoo. And then I smoothly say "yes, a giraffe, I sensed it was a large zoo animal," for which many people will give me full credit—even though no-one would ever mistake a giraffe for a rhino. Again, my alleged psychic powers work because, as an order of magnitude estimate, I am in touch with somewhere between 10,000 and 100,000 people and if the bizarre Barnum statement I come up with happens with a frequency of 1/10,000, at least someone will say "yes" from the crowd. After that, I use the normal deceptive patter that most clairvoyants learn to convince people of their abilities to pierce through the veil and contact the beyond. And they use to the fullest extent the idea that their audience remembers only the successes, rather than the failures. If you knew that already, you must be an Aries.

Devious hucksters aside, orders-of-magnitude estimates are the lifeblood of science and engineering. The poster children for the

---

[2] For more such examples, see Bart Holland's delightful *Voodoo Deaths, Office Gossip, and other Adventures in Probability* (Baltimore, MD: The Johns Hopkins University Press, 2002).

orders-of-magnitude estimates are fluid dynamics people, who turn them into an art form. The Navier–Stokes equation, which describes how fluids move, is over 200 years old. It's nasty to solve, and so usually you resort to simplifications or else to a computer. Suppose you are interested in public health and wonder how dust, or microorganisms, settle in a room; or perhaps you are a cricket or baseball designer wanting to know more about the basic mechanics of how a ball moves through the air. Chances are, you wouldn't want to solve the Navier–Stokes equation, it's too messy. But you do know what force is exerted on the sphere—whether raindrop, dust particle, bacterium, or ball—provided you know the Reynolds number. This is the ratio of the inertial forces to viscous forces but, for our purposes, it's an in-built order-of-magnitude estimator. If a sphere has a diameter D and moves in a fluid of viscosity μ (the Greek letter mu) and density ρ (the Greek letter rho) at speed V, then the Reynolds number Re is $Re = VD\rho/\mu$. If you plug in the numbers for a dust particle, the Reynolds number is tiny. And that means the drag force is proportional to the speed of the particle, something known as Stokes' Law. For a cricket or baseball, launched at speeds of 100 km/h, the Reynolds number is over 100,000, which means that the drag force is quadratic in the velocity and not, like Stokes' Law, linear. This makes life far simpler, as you can get away with solving Newton's second law for those two different cases. In between, it's a mess. In the days of the Soviet Union, the Soviets had one formula to approximate the drag force for intermediate values of the Reynolds number; the United States had another.[3] Take your pick regarding an approximate formula for drag—Communist or Capitalist!

The world of physics brings us Fermi Problems, named after the Italian American physicist Enrico Fermi. The idea is to work out a believable solution to a problem by a combination of logical reasoning and simple numbers. The most famous Fermi Problem, arguably, is the question "How many piano tuners are there in New York City?"

---

[3] Farid Abraham, "Functional dependence of drag coefficient of a sphere on Reynolds number," *The Physics of Fluids*, 13(2194) (1970) has $C = K[1 + \frac{\delta}{\sqrt{Re}}]^2$ with $K\delta^2 = 24$ and $\delta = 9.06$. Klyachko, L. Otopl. I ventil., No. 4 (1934) has $\frac{24}{Re}\left(1 + \frac{Re^{\frac{2}{3}}}{6}\right)$, cited in Ali K. Oskouie, Hwa-Chi Wang, Rashid Mavliev, and Kenneth E. Noll, "Calculated calibration curves for particle size determination based on time-of-flight (TOF)," *Aerosol Science and Technology*, 29(5) (1998), 433–41.

The answer goes in stages, but if you know how long it takes to tune a piano, then you know how many pianos a tuner could tune in a 40-hour work week. Assume that if there were a dearth of tuners, more of them would move into the Big Apple, and if there were too many, some would either leave the city or the profession. Now all you need is to guess what percentage of NYC's 9 million residents actually have a piano, and you have an estimate for how many tuners there are. A far lower percentage than when Fermi posed the problem, I suspect.

Viewed differently, you don't need to know how long it takes to tune a piano, if you know how much a piano tuning cost. Estimate a living wage in NYC, divide by the price of a piano tuning, and that's an order-of-magnitude estimate for how many pianos an expert can tune in a year. Then follow the previous line of logic.

Another Fermi Problem comes from a medical observation, that human temperature isn't the 98.6°F that we thought for decades, it's more like 97.5. So if we could heat the world's 7 billion humans back up to 98.6°F, would it cool down the atmosphere? Fermi Problems are so popular, they even have their own regular column in a physics journal.[4]

An other-worldly example of the use of orders-of-magnitude estimates is the Drake equation, which seeks to calculate the number of extraterrestrial civilizations that can contact us. Once radio telescopes became big enough and sensitive enough to pick up signals from extraterrestrials (assuming they exist), the question arose as to how and whether to look for them. A meeting was organized to discuss the topic and, prior to the meeting, Frank Drake came up with the equation that now bears his name, and it follows a simple chain of logic. First of all, restrict attention to our own galaxy. Next, you need to have a star if you're going to exist, so we need $R(s)$, the rate at which stars are formed in our galaxy per year. Not all of those stars will have planets, so the next ingredient is $f(p)$, the fraction of stars that actually have planets. There's a problem: just because a planet exists doesn't mean it's hospitable to life. In our own solar system, for example, Mercury couldn't, nor— with apologies to Pluto fans—could Pluto (even though we now don't label it as a planet any more). So we also need to incorporate $n(e)$, the number of planets that could support life in any given solar system. We leave it to the philosophers to identify precisely what is meant by life.

---

[4] Fermi Questions, edited by Larry Weinstein, is a regular section in *The Physics Teacher*.

Just because life is possible doesn't mean life will necessarily develop. Life might be possible on Mars, especially if terraforming becomes a thing, but it hasn't yet. So we build in $f(l)$, the fraction of those planets that could support life and that actually then go on and support it. Next, a planet of three-toed sloths would be cute, but might never gain the intellectual horsepower to do much of anything. So of all the planets that support life, only a fraction $f(i)$ go on to support intelligent life.

We're almost done. Of planets that develop intelligent life, only a fraction $f(c)$ produce a civilization that can generate signals that we might be able to detect. Last, and perhaps a chilling reminder to those of us who walk planet Earth, civilizations are not immortal. Call L the length of time that a civilization exists and sends such signals out into space. Guglielmo Marconi sent the first radio message in Italy in 1895, so we have been sending detectable signals into space for just over 100 years and, as a consequence, we could be detected only by civilizations that are within 100 light years of Earth. That includes about 800 stars.

These expressions give us the Drake equation, which says that the number of civilizations in our galaxy with which communication might be possible is:

$$N = R(s)f(p)n(e)f(l)f(i)f(c)L$$

Things turn ugly, though, in estimating each of the parameters. Whether they are true orders-of-magnitude estimates, or simply SWAG, is open to debate. Current computer estimates can differ by a factor of 100, and human calculations range from $N \sim 9 \times 10^{-13}$ all the way up to $N \sim 1.5 \times 10^7$ from zero to oodles.

Unfortunately, the range of parameters that remain open is so wide that you can choose the universe you'd like to live in, a sure example of confirmation bias. Choose small numbers, and we must certainly be alone. Choose large numbers, and we have to ponder why extraterrestrials haven't been in touch. As the old adage goes, either space aliens don't exist, or they are already here!

8

Multiplying and Dividing Irrationally

Beware the irrational, however seductive.
CHRISTOPHER HITCHENS, Letter to a Young Contrarian

There is, in fact, no formal difference between inability to define and stupidity.
ROBERT M. PIRSIG, *Zen and the Art of Motorcycle Maintenance: An Inquiry into Values*

The story so far has been one of integers, added, subtracted, divided, multiplied. To be sure, there have been some quotients and decimal points thrown in occasionally, just to liven things up and set the mathematical pulse racing. The world of numbers, though, includes not just whole numbers, decimals, and fractions. It also includes irrational numbers, ones whose decimal expansions never end, and which crop up almost eerily in all sorts of applications. This chapter focuses on the poster children of irrational numbers: the square roots of two and three; the Golden Ratio; e; and π. The methods curated here are more for mathematical flourish than of practical value. They serve a greater purpose: showing how we can approximate these numbers to get accurate estimates can lead to readers being able to spot their own short cuts for other numbers. Enjoy!

Multiply or divide by $\sqrt{2}$

The Torah, the five books of Moses, states that a Mishkan (a tabernacle) has a courtyard whose area is 50×100 Amos^2. The Mishna, a written compilation of Jewish oral traditions, described the courtyard as having the same area as a square whose side was a little more than 70 Amos. Writing in the twelfth century, the Jewish scholar Maimonides (author of the classic work *A Guide for the Perplexed*) went further, and said that the extra amount was 5/7. This allows us, in an ahistorical way, to come up with a value for $\sqrt{2}$.[1] What Maimonides is telling us is that the area of the courtyard, 5,000, is the square of 70 5/7. Put differently:

$$70\frac{5}{7} = \sqrt{5,000} = \sqrt{\frac{10,000}{2}} = \frac{100}{\sqrt{2}} = \frac{100\sqrt{2}}{\sqrt{2}\sqrt{2}} = 50\sqrt{2}$$

Straight away, we have the approximation in fraction form:

$$\sqrt{2} = \frac{7}{5} + \frac{1}{70} = 1\frac{29}{70}$$

Recalling the short cut from the section "Multiply or divide by 7," in Chapter 4, $5/7 = 0.7142857$. Thus, in decimal form:

$$70.7142857 = 50\sqrt{2}$$

Quick division by 5 (as in, doubling and putting decimal points in the right place) gives a Maimonides-inspired value of $\sqrt{2} = 1.4142857$, close to the actual decimal expansion, which begins 1.4142135, so our answer is extremely close, being about 0.005% lower than the calculator-generated value.

$\sqrt{2}$ shows up in rather a variety of computations in physics, mathematics, and engineering. It is an irrational number, as Maimonides claimed.[2] Its decimal expansion begins $\sqrt{2} = 1.41421356237\ldots$, and this is key to some quick approximate calculations. Namely if we use $\sqrt{2} = 1.4142135$, we are off by less than $5 \times 10^{-6}\%$. This is useful, because:

$$1.4142135 = 7 \times (0.2020305)$$

To compute $8\sqrt{2}$, say, we can immediately multiply by 8 to get:

$$8\sqrt{2} = 7 \times (1.6162440)$$

Multiplication by 7, in this case, is straightforward and gives $8\sqrt{2} = 11.313708$. Alternatively, you could set $7 = 5 + 2$ and then do two

---

[1] For an extended discussion, see Sheldon Epstein, Yonah Wilamowsky, and Bernard Dockman, "Learning Mathematics," *Hakirah: The Flatbush Journal of Jewish Law and Thought*, 14 (2012), 123–39.

[2] There is a delightful proof, by contradiction. If $\sqrt{2}$ is a rational number, we can express it in the lowest possible terms as $\sqrt{2} = m/n$. Squaring, $m^2 = 2n^2$. Hence, m must be even; call it $2k$. But that means $n^2 = 2k^2$. The same logic demands that n is even, say $2p$. However, that contradicts the assumption that m/n is in the lowest form possible (k/p would be). Hence, there is a contradiction and $\sqrt{2}$ is irrational, as it can't be expressed by such a fraction.

separate multiplications. The actual answer is 11.3137085 . . . As far as division goes, break out the standard trick of multiplying by 1, which in this case takes the form of $\sqrt{2}/\sqrt{2}$. To find $8/\sqrt{2}$, multiply top and bottom to get $8\sqrt{2}/2$. You calculate $8\sqrt{2}$ as shown above and halve the answer. Easy!

Multiply or divide by $\sqrt{3}$

In elementary trigonometry, herds of $1, \sqrt{3}, 2$ right-angled triangles sweep majestically across the mathematical planes. But how to multiply and divide by $\sqrt{3}$, given that it crops up so often in the sines and cosines of angles? As with $\sqrt{2}$, division is easy once we know how to multiply.

Here we use the technique for a number whose digits are multiples of each other. Namely, as $\sqrt{3} = 1.732050808$, we'll strip it down to the far easier to handle, and quicker to calculate, 1.732, accurate to 0.003%, and a touch on the low side. Naturally, what we'll do is write this as 1.632 and then add on 0.1. To calculate $27\sqrt{3}$, we know that $27 \times 16 = 432$. We expect an answer of about 40. So our first guess is 43.2. Double this to get 86.4, shunt over two places to form 0.864 and add to the original 43.2 to give 44.064. Now we need to mix in 0.1 of the original number, which is 2.7. Thus, $27\sqrt{3} = 44.064 + 2.7 = 46.764$, which is within a gnat's breath of the calculator value, 46.765371.

Multiply or divide by the Golden Ratio

There is much nonsense written about the Golden Ratio, φ, and there is a cottage industry of "Golden Ratio" spotters who claim that something they have discovered looks pretty, precisely because the dimensions are in a Golden Ratio. Let's focus, instead, on the math. Euclid was the first to explore the properties of a line, of length $a + b$, that was divided into two parts: perhaps obviously, one of length a, the other of length b. Suppose $a > b$. If the ratio of b to a is the same as the ratio of the entire line to length a, then the two parts lie in a Golden Ratio to each other. To focus on the arithmetic, two numbers a and b, where both are positive and b is larger than a, are in a Golden Ratio if:

$$\frac{a + b}{a} = 1 + \frac{b}{a} = \frac{a}{b}$$

The Golden Ratio, φ, is defined by:

$$1 + \frac{1}{\varphi} = \varphi$$

So that:

$$\varphi^2 - \varphi - 1 = 0$$

For those who know the quadratic formula (and even for those who don't!) the positive solution to this equation is:

$$\varphi = \frac{1 + \sqrt{5}}{2} = 1.618033988\ldots$$

The Golden Ratio is one of a select club of numbers that merit their own book.[3] In fact, it has more than one book devoted to it. Luca Pacioli wrote *De Divina Proportione* (*On the Divine Proportion*) about the Golden Ratio. Written around 1498 but published in 1509, it contained diagrams by Leonardo da Vinci.[4]

Apart from the sheer joy of being able to compute multiples of the Golden Ratio swiftly, the great advantage of considering such multiples is to build up an ability to spot your own quicker-calculation formulas—the challenge being, "If I had to calculate a multiple of the Golden Ratio, how could I do it?" One way is to see that $\varphi \approx 1.618$. This involves the numbers 16 and 18, both of which are either side of 17. To calculate 7φ, say, you calculate $7 \times 17 = 119$. Add another 7 to get $7 \times 18 = 126$ and subtract a 7 to get $7 \times 16 = 112$. Shunt the 126 over two places to the right (or add a couple of zeros to the end of 112) to obtain 11,326. As the answer is about 10, we have $7\varphi \approx 11.326$.

And if you want to divide by the Golden Ratio? Not a problem! Simply shuffle the equation

$$1 + \frac{1}{\varphi} = \varphi$$

around to get:

$$\frac{1}{\varphi} = \varphi - 1$$

To find $7/\varphi$, simply find 7φ and subtract 7.

---

[3] Mario Livio, *The Golden Ratio: The Story of Phi, The World's Most Astonishing Number* (New York: Broadway Books, 2002).

[4] Johannes Kepler, of Kepler's three laws of planetary motion fame, also wrote about the number. He called it the "sectio divina," the divine section. This might be a groanworthy Latin pun on "lectio divina," the reading that Catholic priests and nuns have to do each day.

Multiply or divide by e

The base of natural logarithms, $e = 2.718281828$ (also known as Euler's number, after Swiss mathematician Leonhard Euler [1707–83]) is so famous it has its own biography, Eli Maor's *e: The Story of a Number*. While no-one is typically asked to calculate multiples of e in a hurry, there is a fairly simple way of finding approximate answers swiftly, which can impress the mathematically keen and eager. The first method uses the approximation $e \approx 2.75$, which is slightly too high, but is accurate to within 1.17%. The way to use this is to note (or to know!) that 2.75 is the average of 2.5 and 3. Hence, if you seek ne for some number n, compute $25n$, add this to $30n$, and halve the answer. Then insert a decimal point in the right place, and you're done.

Consider $14e$. Start by working out 25×14. We know (multiply by 15, 25, . . .) to divide by 4. So 14 when halved is 7 and this, when halved, is 3.5. Our answer, then, is $25 \times 14 = 350$. Now $3 \times 14 = 42$, so $30 \times 14 = 420$. Add the two approximations together, which leads to $350 + 420 = 770$. To find the average, halve the answer, which is 385. As the answer has to be less than $14 \times 3 = 42$, we have $14e = 38.5$. The actual answer is 38.0559456. . . .

But we can dazzle mathematics enthusiasts with a blisteringly accurate approximation. Replace $e = 2.718281828 \ldots$ by 2.71827, a slight underestimate that is accurate to better than 0.0044%. The key is to see that:

$$2.71827 = 9 \times (0.30203)$$

And so the only calculations we need to do are to double and treble the number, and be able to multiply by 9.

Take $14e$ as the example once again: $14 \times 2 = 28$, while $14 \times 3 = 42$. Thus we have:

$$14e \approx 9 \times (4.22842)$$

And this is $42.2842 - 4.22842 = 38.05578$, which agrees with the actual answer up to, and including, the third decimal place (remember: if you don't like subtraction, you can multiply 4.22842 by 3, and then do so again).

To divide by e, we can't get quite so close. But $1/e = 0.367879441 \ldots$. We can take a stab at the answer by writing this as approximately $0.36 + 0.0072 + 0.00036 = 0.36756$, which is good to within 0.09%. All we need to do, then is to multiply by 36, double, and shunt a few times!

To get 14/*e*, we take the 14 and multiply by 12 to get 168. Now multiply by 3 to get 504. That's the multiplication by 36 over and done with. We double it to get 1,008. Then we write down:

5.04
0.1008
0.00504

And sum to get 5.14584

By now, as you may suspect, we can do better. If we added on a further 0.00504, we get 5.15088, which is within a hair's breadth of the computed answer of 5.150312176... The reason for this success is that it has changed our approximation from 0.36756 to 0.36756 + 0.00036 = 0.36792. This is within 0.012% of the actual answer.

Multiply or divide by π

And he made a molten sea, ten cubits from the one brim to the other: it was round all about, and his height was five cubits: and a line of thirty cubits did compass it round about.

1 Kings 7:23 KJV seemingly using the approximation $\pi = 3$

...The ratio of the diameter and circumference is as five-fourths to four.

Indiana House Bill No. 246, 1897, which was defeated, attempting to legislate that $\pi = \frac{16}{5} = 3.2$

The ratio of the circumference of a circle to its diameter is called by the Greek letter pi, π. This choice, though, is another mathematical invention, together with the equals sign, for which the Welsh can take credit. William Jones (1675–1749), a friend of Sir Isaac Newton and of Sir Edmund Halley (of Halley's Comet fame), first used it in his book *Synopsis Palmariorum Matheseos* (also called *A New Introduction to the Mathematics*), published in 1706. Jones, who hailed from the village of Benllech on Ynys Môn, also known as Anglesey, gains the glory of introducing the overdot into differential calculus, used to denote the change of a quantity with time.

But what is the value of π? If you struggle with remembering the first few digits (3.14 is easy), none other than famous British astronomer

Sir James Jeans comes to the rescue: "How I want a drink, alcoholic of course, after the heavy lectures involving quantum mechanics" was his way to remember the first 15 digits—just note the number of letters in each word![5] If you want to go a tad farther, you could add "now we owe Einstein four liters of frothy beer" to get the next nine digits. There are folks who memorize the digits of pi, some of them can rattle off the first 100,000, and YouTube features a "Pi song" to help you learn the first 100, if that's something you'd like to do.

In the ancient world, there were a number of approximations. One obvious approximation, used in Babylon and possibly by the Jewish people, is 3. The Babylonians also used 25/8, which is 3.125, accurate to 0.6%, and as it involves multiplication by 25 and division by 8, is a great form to use for quick approximations.

Archimedes posited that π was less than the well-known high-school approximation, 22/7 ($= 3.142857\ldots$), which is an overestimate accurate to 0.4%. There are more accurate fraction approximations known in China, say, or decimal expansions, and the story of π and approximations to it deserves to be more widely known (see Petr Beckman's *A History of Pi*). Here, though, we introduce a method to write down multiples of π that is swift and extraordinarily accurate.

A slice of Babylonian π

Should you wish to calculate 6π, use the Babylonian approximation and write this as $6 \times 25/8 = (6 \times 3) + 6/8$. Hence, we write down $6\pi = 18.75$. For 8π, we can write $8\pi = 24 + 1 = 25$.

This hints at how to form multiples of π in your head. If you are asked for a multiple of 8, use the Babylonian approximation; if you are asked for a multiple of 7, use the 22/7 approximation.

As an aside, here's a fun formula that follows from Archimedes' approximation. If you know multiple-angle formulas, recall that:

$$\sin^2 A = \frac{1}{2}[1 - \cos 2A]$$

Set $A = 11$ radians to obtain:

$$\sin^2 11 = \frac{1}{2}[1 - \cos 22]$$

---

[5] As reported in Alan S. Hawkesworth, "Two Mnemonics," *The American Mathematical Monthly*, 38(3) (1931), 158.

Archimedes' approximation leads to:

$$\sin^2 11 \approx \frac{1}{2}[1 - \cos 7\pi] = 1$$

Leading to:

$$\sin 11 = -1$$

There are other fractions that are close to π; for example, 333/106 (good to four digits), 355/113 (good to six digits), and 1,442/459. The last of these three (although only good to three digits but more accurate than 333/206) has an even numerator, so we can perform the same trick. Namely, set $A = 721$ radians to get:

$$\sin^2 721 = \frac{1}{2}[1 - \cos 1,442] \approx \frac{1}{2}[1 - \cos 459\pi]$$

Consequently, $\sin 721 = -1$. Go ahead and show that $\sin(355)$ and $\sin(166.5)$ are approximately 0.

Similar, but different

We can improve our approximation by using 3.15 instead of 3.125, as this will be accurate to within 0.27% of the actual value. By now, ignoring the decimal points should be habitual. If we seek the number $n\pi$, we form $315n$, which we can write as $315n = 300n + 15n = 100(3n) + 5(3n)$, and this is relatively simple. Consider 12π. Guesstimating (or, to be more formal, using the Book of Kings approximation of 3^6) gives an answer of 36. That is to say, we set $\pi = 3$ to get $12 \times 3 = 36$ and then affix two zeros to get $100(3n) = 3,600$. Next, multiply 36 by 5, which we do by dividing 36 by 2, to get 18, and strapping on a zero to get 180. Our approximation, then, is $3,600 + 180 = 3,780$, but with a decimal point inserted so that the value is about 36. Thus $12\pi \approx 37.8$, which is close to the more accurate answer of 37.69911. . .

---

[6] This is perhaps unfair to the scribes of the Book of Kings. Three verses later, the Bible says the rim of the bowl was a hand's breadth wide. Medieval Jewish scholar Levi ben Gershon reconciled the biblical account with mathematics by saying that if the circumference is measured along the outside of the bowl, but the diameter is measured on the inner side, the ratio of measurements becomes $\pi = 90/29 = 3.1034$. see Shai Simonson, "Mathematics of Levi ben Gershon in the classroom," *Convergence*, Mathematical Association of America. https://www.maa.org/press/periodicals/convergence/the-mathematics-of-levi-ben-gershon-in-the-classroom (accessed November 6, 2020).

Also similar, but far more accurate

We can home in on multiples of π by using the approximation 3.14, which is accurate to better than 0.051%. As before, we concern ourselves only with $314n = 300n + 14n$. Let's work out the same example. For 12π, we know $n = 12$, which we triple to get 36. Hence $300n = 3,600$. Now we take the 12, multiply by 7 to get 84, double to get 168. Adding the two gives us an approximation of $12\pi = 37.68$, after inserting a decimal point.

Given that we now know how to multiply rapidly by 3.14 and 3.15, it isn't that hard to multiply by 3.16. This leads to a worse approximation for π, admittedly, but this value was used in Problem 41 of the Rhind papyrus, a mathematical text from Ancient Egypt (now housed in the British Museum), which dates close to 1550 BCE.

The approximation 3.14 leaves a slight shortfall compared with the actual answer, but that allows us to end with a flourish. Namely, we know that our answer is 37.68, but if we take the 168 that we calculated and divide it by 100 and add, we get 37.6968. What we are doing, naturally, is to replace π by 3.1414, which is accurate to better than 0.009% of the actual answer, 37.69911184. . . What if you prefer a different flourish? To compute 12π, look at the 12. To begin, multiply it by 3 to get 36. Multiply this number by 5 to obtain 180. Subtract the original 12 to get 168 and divide by 100 to get 1.68. Now add the original 12 to the 180 to get 192, but now shunt four places to the right, to get 0.0192. Add them up: $36 + 1.68 + 0.0192 = 37.6992$.

This is beautifully short and impressively accurate. By multiplying first by 3 we obtain $3n$. We multiply by 5 to get $15n$, subtract n and divide by 100 to get $0.14n$. We add n and divide by 10,000 to get $0.0016n$, and hence we approximate π by 3.1416, accurate to within 0.00023%.

An almost-precise method

To evaluate 6π, first write down $6 \times 3 = 18$. Next, take the number 6 and multiply by 7 to get 42; double this to obtain 84. To get our answer, you need to shunt this 84 over a few places, and add it several times. To be more specific, write it as 0.84, then shunt it over two places, three places, and five places (twice!) and then add. You'll also need to add half of the multiple, which as we seek 6π means we need to add in a 3 (actually, it's better to think of this as a 30). Write down 0.3 and shunt it over three places. This sounds vague![7] For clarity, you write down the sum:

$$18.0000000$$
$$.8400000$$
$$.0084000$$
$$.0008400$$
$$.0003000$$
$$.0000084$$
$$\underline{.0000084}$$
$$18.8495568$$

The bold in the top line shows how to align the *last* digits of the various "84" entries and the "30" entry.

This is extremely close to, but slightly above, the actual answer, which is 18.8495559. . . A trickier example that works on the same basis, but which is more awkward because the numbers are more than 100, is 8π. Start as before with $3 \times 8 = 24$. As $8 \times 7 = 56$, which when doubled gives 112, the sum now is:

$$24.000000$$
$$1.12$$
$$.0112$$
$$.00112$$
$$.00040$$
$$.0000112$$
$$\underline{.0000112}$$
$$25.1327424$$

Whereas $8\pi = 25.13274123 \ldots$

While this may look or sound complicated, with a mere smidgeon of practice you can write down the answer, accurate to within thousandths of a percent, in less time than it takes to go get a calculator.

[7] For the same method, but explained using different words, see Trevor Lipscombe, "Mental mathematics for multiples of π," *Mathematical Gazette*, 96(1) (2013), 72–4.

How it works

By multiplying the number by 7 and doubling it, we are clearly doing a rapid multiplication by 14. The key is in the third line, in which we have 0.0003, which is half of the multiple, 6, with decimal places added. What we are doing, then, is to construct $3 + 0.1414 + 0.00014 + 0.00005 + 0.0000028 = 3.1415928$, which differs from $\pi = 3.141592654\ldots$ in the seventh decimal place and is accurate to within $5 \times 10^{-6}\%$.

Divide by π

To divide by π, note that $1/\pi = 0.31831\ldots$ which is approximately 0.31830, or $0.30 + 0.018 + 0.0003$, accurate to within 0.0032%. All you need to do is to multiply the integer by 3, multiply the result by 6, and you have all the numbers needed to get your answer!

To find $7/\pi$, multiply $7 \times 3 = 21$ and insert a decimal point to get 2.1. Multiply the 21 by 6 to get 126 and insert the decimal point to get 0.126, so the total is $7/\pi = 2.226$. Now add on the last multiplication by 3, which gives $7/\pi = 2.2281$. The calculator answer is 2.228169203...

For those who like mathematical mysteries (and who doesn't?) there are a couple of fascinating "almost" formulas that include both e and π. It turns out that $e^{\pi} - \pi = 19.999099979\ldots$ and $\pi^9/e^8 = 9.9998387\ldots$, which are known as "almost integers." If you can discover why this is so, mathematical fame and glory may well await.

Try these

1. $0.32e$
2. $4.3/\pi$
3. 0.11φ
4. $0.64/\sqrt{2}$
5. $4.2\sqrt{3}$
6. $5.7/e$
7. 0.24π
8. $3.6/\varphi$
9. $0.37\sqrt{2}$
10. $3.3/\sqrt{3}$
11. $75e$
12. $42/\pi$
13. 58φ

14. $2.1/\sqrt{2}$
15. $850\sqrt{3}$
16. $770/e$
17. 8.9π
18. $510/\varphi$
19. $150\sqrt{2}$
20. $0.31/\sqrt{3}$
21. $0.079e$
22. $42/\pi$
23. 0.74φ
24. $140/\sqrt{2}$
25. $18\sqrt{3}$
26. $7.1/e$
27. 180π
28. $440/\varphi$
29. $6.7\sqrt{2}$
30. $840/\sqrt{3}$
31. $0.69e$
32. $3.1/\pi$
33. 48φ
34. $99/\sqrt{2}$
35. $0.42\sqrt{3}$
36. $420e$
37. $7.8/\pi$
38. 0.16φ
39. $0.15/\sqrt{2}$
40. $0.02525\sqrt{3}$
41. $27/e$
42. 750π
43. $6/\varphi$
44. $20\sqrt{2}$
45. $15/\sqrt{3}$
46. $0.084e$
47. $9.2/\pi$
48. $0.34/\varphi$
49. $0.95\sqrt{2}$
50. $55/\sqrt{3}$

The Grand Finale

The (not so) Grand Finale: Go back to the Challenge posed in the front matter of the book, and see how long it takes you to compute the answers!

Further Reading

Of making many books *there is* no end; and much study *is* a weariness of the flesh.

Ecclesiastes 12:12 (King James Version)

Other works on rapid mathematics

Benjamin, Arthur and Michael Shermer, *Secrets of Mental Math: The Mathematician's Guide to Lightning Calculation and Amazing Math Tricks* (New York: Three Rivers Press, 2006).

Collins, A. Frederick, *Rapid Math without a Calculator: Shortcuts for Mastering Fast Addition, Subtraction, Division, Multiplication, Fractions and More!* (New York: Citadel Press, 1987).

Julius, Edward H., *Rapid Math Tricks and Tips: 30 Days to Number Power* (New York: John Wiley and Sons, 1992).

Julius, Edward H., *Rapid Math in 10 Days: The Quick-and-Easy Program for Mastering Numbers* (New York: Perigee Trade, 1994).

Julius, Edward H., *Arithmetricks: 50 Ways to Add, Subtract, Multiply, and Divide without a Calculator* (New York: John Wiley and Sons, 1995).

Julius, Edward H., *More Rapid Math Tricks and Tips: 30 Days to Number Mastery* (New York: John Wiley and Sons, 1996).

Kelly, Gerard W., *Short-Cut Math* (New York: Dover Books, 1984).

Trachtenberg, Jakow, translated and adapted by Ann Cutler and Rudolph McShane, *The Trachtenberg Speed System of Basic Mathematics* (New York: Ishi Press, 2011).

Numbers

Beckmann, Petr, *A History of π* (Golem Press: New York, 1971).

Kaplan, Robert, *The Nothing That Is: A Natural History of Zero* (Oxford: Oxford University Press, 1999).

Livio, Mario, *The Golden Ratio: The Story of Phi, the World's Most Astonishing Number* (New York: Broadway Books, 2002).

Maor, Eli, *e: The Story of a Number* (Princeton, NJ: Princeton University Press, 1993).

Maor, Eli, "Eleven: The first uninteresting number?," *Mathematics Teaching in the Middle School*, 7(5) (2002), 308–11.

Nahin, Paul, *An Imaginary Tale: The Story of $\sqrt{-1}$* (Princeton, NJ: Princeton University Press, 1998).

Seife, Charles, *Zero: The Biography of a Dangerous Idea* (New York: Penguin, 2000).

History of mathematics

Acheson, David, *1089 and All That: A Journey into Mathematics* (Oxford: Oxford University Press, 2002).

Asher, Marcia, *Mathematics Elsewhere: An Exploration of Ideas across Cultures* (Princeton, NJ: Princeton University Press, 2002).

Gazale, Midhat, *Number: From Ahmes to Camtor* (Princeton, NJ: Princeton University Press, 2000).

Joseph, George Gerveghese, Crest of the Peacock: *Non-European Roots of Mathematics* (New York: I.B. Tauris, 1991).

Kline, Morris, *Mathematical Thought from Ancient to Modern Times* (Oxford: Oxford University Press, 1972). Three-volume paperback edition, 1990.

Lo Bello, Anthony, *Origins of Mathematical Words: A Comprehensive Dictionary of Latin, Greek, and Arabic Roots* (Baltimore, MD: The Johns Hopkins University Press, 2013).

Shell-Gellasch, Amy and John Thoo, *Algebra in Context: Introductory Algebra from Origins to Applications* (Baltimore, MD: The Johns Hopkins University Press, 2015).

Swetz, Frank, *Mathematical Expeditions: Exploring Word Problems across the Ages* (Baltimore, MD: The Johns Hopkins University Press, 2012).

Calculating Doomsday

The Doomsday Interlude showed how to come up with the day of the week for any date that someone asks. The key, though, is to know what the Doomsday for any given year happens to be. It's easy enough for the current year, as you can simply find June 6 in a calendar and whatever day of the week it is will be the Doomsday for that year. But we can do better. The Doomsday algorithm is complicated, but there is a simple expression for years in the range of 1900–99 and from 2000–99.

You need to calculate:

$$\left[\frac{Year\left(last\ two\ digits\right) + \left\lfloor \frac{Year\left(last\ two\ digits\right)}{4} \right\rfloor + Day\ within\ month + Monthly\ factor}{7} \right]$$

Here, $\lfloor \ldots \rfloor$ is the floor function, which means you do the calculation, then round all the way *down* to the nearest integer. The integer remainder, when you divide by 7, lets you know the specific day of the week.

This is the expression you need for years in the range 1900–99. For those in the 2000–99 period, the same expression holds, except that you need to subtract 1 from the numerator before dividing by 7.

Archimedes allegedly said "There is no royal road to geometry." If you want to learn something, you don't get a free pass because your mother is a marchioness or your aunt an aristocrat. Likewise, to do a swift calendrical calculation, there are no short cuts. The monthly factors are: January 1 (0 for a leap year); February 4 (3 for a leap year); March 4; April 0; May 2; June 5; July 0; August 3; September 6; October 1; November 4; December 6.

A harmless ditty helps to remember the month factors:

> July and April: zero
>
> Jan and Oct are one
>
> May alone has factor two
>
> (Which is a lot of fun)
>
> August has the factor three
>
> Feb, March, November: four
>
> June alone has factor five
>
> (A fun math fact for sure)

September and December,

Both of these are six

Behold the monthly factors

For date-related tricks.

Luckily, the remainder is straightforward. Sunday is 1, Monday is 2, Tuesday is 3, and so on up to Friday, which is 6, and then Saturday, which is 0.

As an example. What day of the week was July 19, 1969, the day that humans first stood on the moon? The two-digit version of the year is 69. Divide this by 4 to get 17.25. Applying the floor function, this becomes 17. We then get, as July's monthly factor is zero:

$$\frac{\lfloor 69 + 17 + 19 \rfloor}{7} = \frac{105}{7}$$

This has a remainder of 0 and so the day of the week was a Saturday.

A bigger question is to work out Doomsday for a year. For this, choose the month of April, as the month factor for April is 0. We know that 4/4 always falls on Doomsday. Hence, for any year in the 1900s, we have:

$$\frac{\left\lfloor Year\left(last\ two\ digits\right) + \left\lfloor \frac{Year(last\ two\ digits)}{4} \right\rfloor + 4 \right\rfloor}{7}$$

For 1945, the year that the Second World War came to an end, Doomsday is given by:

$$\frac{\lfloor 45 + 11 + 4 \rfloor}{7} = \frac{60}{7}$$

We care only about the remainder, which is 4, so that Doomsday, 1945 was Wednesday.

For the twenty-first century, Doomsday is:

$$\frac{\left\lfloor Year\left(last\ two\ digits\right) + \left\lfloor \frac{Year(last\ two\ digits)}{4} \right\rfloor + 3 \right\rfloor}{7}$$

For 2020, the year of global pandemic, Doomsday was:

$$\frac{\lfloor 20 + 5 + 3 \rfloor}{7} = \frac{28}{7}$$

Which has a remainder of 2, and its Doomsday was Saturday.

As you've seen, for the twenty-first century, add 4 in the numerator; for the twentieth century, add 3. To figure out Doomsday for the nineteenth century, add 6 (or take off 1), and for the eighteenth century, add on 1.

Maria Gaetana Agnesi wrote the first book on integral and differential calculus. She was born in Milan on May 16, 1718. African American surveyor Benjamin Banneker came into the world on November 8, 1731. The famous Indian mathematician Ramanujan was born on December 22, 1887. Deaf-blind Dutch mathematician Gerrit van der Mey was born on January 5, 1914. And African American Katherine Johnson, who researched orbital mechanics at NASA and who became famous through the book and movie *Hidden Figures*, was born on August 26, 1918. To celebrate their accomplishments, work out the days of the week on which they were born.

APPENDIX II

The Squares from 1 to 100

It is not so very important for a person to learn facts. For that, he does not really need college. He can learn them from books. The value of an education in a liberal arts college is not the learning of many facts, but the training of the mind to think something that cannot be learned from textbooks.

Albert Einstein. Quoted in Philipp Frank, *Einstein: His Life and Times*, page 185

In spite of what Einstein said, it can sometimes help to have a store of facts laid up for when you need them. The squares of integers can be useful to know. However, one way to check what you have picked up from this book is to see how quickly you can write all of these down—before you memorize them—by using the methods you have picked up from these pages. But, to concur with Einstein, the real challenge is to train your mind to think of your own short cuts, to make rapid mathematics not so much a topic that you study, but the way you naturally do arithmetic.

$1^2 = 1$

$2^2 = 4$

$3^2 = 9$

$4^2 = 16$

$5^2 = 25$

$6^2 = 36$

$7^2 = 49$

$8^2 = 64$

$9^2 = 81$

$10^2 = 100$

$11^2 = 121$

$12^2 = 144$

$13^2 = 169$

$14^2 = 196$

$15^2 = 225$

$16^2 = 256$

$17^2 = 289$

$18^2 = 324$

$19^2 = 361$

$20^2 = 400$

$21^2 = 441$

$22^2 = 484$

$23^2 = 529$

$24^2 = 576$

$25^2 = 625$

$26^2 = 676$

$27^2 = 729$

$28^2 = 784$

$29^2 = 841$

$30^2 = 900$

$31^2 = 961$

$32^2 = 1,024$

$33^2 = 1,089$

$34^2 = 1,156$

$35^2 = 1,225$

$36^2 = 1,296$

$37^2 = 1,369$

$38^2 = 1,444$

$39^2 = 1,521$

$40^2 = 1,600$

$41^2 = 1,681$

$42^2 = 1,764$

$43^2 = 1,849$

$44^2 = 1,936$

$45^2 = 2,025$

$46^2 = 2,116$

$47^2 = 2,206$

$48^2 = 2,304$

$49^2 = 2,401$

$50^2 = 2,500$

$51^2 = 2,601$

$52^2 = 2,704$

$53^2 = 2,809$

$54^2 = 2,916$

$55^2 = 3,025$

$56^2 = 3,136$

$57^2 = 3,249$

$58^2 = 3,364$

$59^2 = 3,481$

$60^2 = 3,600$

$61^2 = 3,721$

$62^2 = 3,844$

$63^2 = 3,969$

$64^2 = 4,096$

$65^2 = 4,225$

$66^2 = 4,356$

$67^2 = 4,489$

$68^2 = 4,624$

$69^2 = 4,761$

$70^2 = 4,900$

$71^2 = 5,041$

$72^2 = 5,184$

$73^2 = 5,329$

$74^2 = 5,476$

$75^2 = 5,625$

$76^2 = 5,776$

$77^2 = 5,929$

$78^2 = 6,084$

$79^2 = 6,241$

$80^2 = 6,400$

$81^2 = 6,561$

$82^2 = 6,724$

$83^2 = 6,889$

$84^2 = 7,056$

$85^2 = 7,225$

$86^2 = 7,396$

$87^2 = 7,569$

$88^2 = 7,744$

$89^2 = 7,921$

$90^2 = 8,100$

$91^2 = 8,291$

$92^2 = 8,464$

$93^2 = 8,649$

$94^2 = 8,836$

$95^2 = 9,025$

$96^2 = 9,216$

$97^2 = 9,409$

$98^2 = 9,604$

$99^2 = 9,801$

$100^2 = 10,000$

Index

Index

Independence Day
  India 78
  United States 69, 78
India 78, 98–100, 116
Inland Revenue Service 23
ISBN 39, 40
Ishango bone 20

Jeans, Sir James 155
Jefferson, Thomas
  Age 22
  Second 91
  Slavery 105
  Treaty of the Meter 53
Jesus 63
Jinches (a Jeffersonian inch) 91
Johns Hopkins University 26
Johnson, Lyndon 22, 91
Jones, William 154
Juan Diego, Saint 78
Juneteenth (June 19th) 78
Jupiters 64

Kardashian, Kim 79
Katrina and the Waves 80
Kepler, Johannes 15, 139
Kepler Conjecture 139
Kermits (unit of time) 126, 127
KerMetric 126
al-Khwarizmi 127
Kidney Stones 85
King Kong 81

Ibn Labbān, Kūshyār 117
Lattice Method 98–100
Lawsuits, largest settlements 36
Leonardo of Pisa 51, 127, 129
Lerwick, Scotland 56
Lewis, C(live) S(taples) 104
Līlāvatī 98
Lincoln, Abraham 91

Magdalen College, Oxford 104
Magnitude, Order of 46, 130, 145, 146
Magi 63
Maimonides 149, 150
Malmsey 61
Marconi, Guglielmo 148
Marić, Mileva 79

Marmite 67
Mathemagic 13, 14
Mayans (calendrical calculations) 68
Medical schools and MCAT scores 32, 33
Melchior (measure of volume) 63
Mersenne, Marin 43
Mishna 149
Mishkan 149
Molecular weight 62–64
Mongolian People's Republic 114
Moriarty, Professor James 26
Movies, highest grossing 34, 35
Muhammed Ali (BMI calculation) 104

Naeh, Rabbi Avrahim Chaim 45
Nanoputians 64
NASA 84
Navier-Stokes Equation 146
Nelson, Horatio 61, 72
Nertz 68
Newcomb, Simon, and Newcomb-Benford
  Law 26
Number
  Cannonball 138
  Happy 57
  Kaprekar 140
  Kindred 89
  Mirror 10
  Palindrome 14
  Perfect 47
  Powerful 93
  Prime 14, 43, 44, 66, 86–88, 93
  Reynolds 146
  Sphenic 138
Numerals
  Egyptian 102
  Indo-Arabic 97, 98, 102
  Roman 97, 127

Order of Magnitude 46, 130, 145, 146
Oxford 104
Oysters 67

Pacioli, Luca 16, 152
Palaces, property values 35, 36
Papyrus 58, 116, 157
Penguinone 63
Pesci, Joe (BMI calculation) 104
Pfiffig, Herr 14